THE ILLUSTRATED HISTORY OF
NASA

The illustrated history of

NASA

ANNIVERSARY EDITION

NASA

United States

ROBIN KERROD

GALLERY BOOKS
An Imprint of W. H. Smith Publishers Inc.
112 Madison Avenue
New York City 10016

This book was devised and produced by
Multimedia Publications (UK) Ltd

Editor: Jeff Groman
Assistant Editor: Tony Hall
Production: Karen Bromley
Design: Michael Hodson Designs

First published in the United States of
America by

GALLERY BOOKS
An imprint of **W H SMITH PUBLISHERS INC.**,
112 Madison Avenue,
New York, NY 10016

ISBN 0-8317-4878-8

Revised edition 1988

Typeset by Rowlands Typesetting
Origination by Inline Studios, London
Printed in Italy by New Interlitho Spa

Page 1: Astronauts in the
Apollo 11 lunar module took
this picture of the command
and service module (CSM) in
lunar orbit in July 1969.

Title page: The prototype
shuttle orbiter Enterprise rides
piggyback on its Boeing 747
carrier jet during approach and
landing tests in 1978.

This book is dedicated to the memory of the seven US
astronauts who lost their lives on January 28, 1986: Dick
Scobee, Mike Smith, Judy Resnik, Ron McNair, Ellison
Onizuka, Greg Jarvis and Christa McAuliffe.

These pages: A famous picture
that captures the spectacle of
the Apollo missions. James
Irwin salutes the Stars and
Stripes at the Apollo 15 landing
site.

CONTENTS

Introduction

On 1 October 1958 the body we know as NASA, the National Aeronautics and Space Administration, came into being. For more than three decades now NASA has been guiding the American aerospace effort, through triumph and through tragedy, establishing the United States as the pre-eminent nation in aerospace technology and exploration. NASA's meticulous and painstaking planning, development and preparation of hardware and software have over the years insured that the triumphs have been many, the tragedies few. But the dangers inherent in probing new frontiers in the air and in space are ever present and never to be underestimated.

This was brought home to the world at large in a particularly horrifying manner in January 1986 when orbiter *Challenger* exploded in a spectacular fireball just after lift-off. In that fireball seven American astronauts met their death, martyrs to Man's cosmic wanderlust.

As the world mourned the death of seven heroes in the worst disaster of the Space Age, NASA and President Reagan confirmed that the American space program would go ahead as soon as the cause of the disaster became known. Declared the President: 'Other brave Americans must follow where they (*Challenger*'s crew) so valiantly tried to lead ... to pursue their dreams to the stars and beyond.'

NASA was spawned from an organization known as NACA, the National Advisory Committee for Aeronautics. This was founded in 1915 by far-sighted men who were concerned with the then-primitive state of aviation technology in the United States. And NASA is still heavily engaged in research and development work in all aspects of aviation. But to the world at large, NASA means space.

NASA picked up the threads of an embryonic and fragmented space program that had been initiated largely by the military. It was in fact a US Army team under rocket pioneer Wernher von Braun that launched the first American satellite, Explorer 1, on the last day of January 1958. Unbelievably, *within a decade,* three Americans were reading passages from the Book of Genesis while they were orbiting around the Moon!

From a pint-sized satellite in low Earth orbit to a lunar circumnavigation within 10 years was a startling technological leap, made possible by the injection of upward of $25 billion and a single-minded national resolve to beat the Russians to the Moon. By 20 July 1969 the 'race to the Moon' was won, as Neil Armstrong planted an American foot in the dusty lunar soil. The impossible had been achieved.

The Moon-landing program, Project Apollo, generated purpose-built hardware that had no future. Even before the last of the Apollo astronauts had brushed the lunar soil from his boots, the

space transportation system of the future was on the drawing board. Today that system is operational. It is the reusable space shuttle. Today, shuttles carry into space crews of up to seven astronauts, both men and women, and many tons of cargo in their cavernous cargo holds.

The key element in the shuttle system is the tile-covered orbiter, a vehicle that represents a confluence of aeronautic and space technologies, just as NASA itself does. The orbiter takes off like a rocket, spewing fire from its engines. In orbit it maneuvers like a spacecraft by means of jet thrusters. On returning to Earth it transforms from a spacecraft to a glider. It becomes aerodynamic, receiving lift from its delta wings and directional control from its

Above: Critical to the success of the manned spaceflight program is Houston mission control, seen here at the end of the triumphant first flight of the space shuttle. Orbiter Columbia *appears on the video screens at the front of the mission control room. The date is 14 April 1981.*

elevons and tail rudder. And it lands on an ordinary runway like a terrestrial airliner.

Up in space the orbiters demonstrate their versatility on every mission. With the assistance of free-flying human satellites, they act as in-orbit service stations to repair and retrieve ailing satellites. Soon they will be ferrying up components of a space station, which will be assembled in orbit.

Though it is their manned flights that capture the headlines, NASA's achievements in the workaday world of ordinary satellites is also extraordinary. These ingenious man-made moons flash telephone, telex, TV and data signals from continent to continent; they warn of approaching storms; spot new mineral deposits; guide ships across the oceans; and help save pilots' lives

Out-of-this-world NASA space probes have revolutionized many branches of astronomy. They have flown by Venus, reporting a hellish world; landed on Mars and tested the soil for sign of life; spotted volcanoes erupting on Jupiter's moon Io; and measured winds on Saturn traveling at the speed of a rifle bullet!

After scanning the planets in our solar system, the Pioneer and Voyager probes will eventually disappear into interstellar space, perhaps to encounter other planets of other stars, thousands or hundreds of thousands of years hence. Then perhaps intelligent extraterrestrials may find them and learn of beings that lived on a planet called Earth and of a body that called itself NASA.

Chapter 1

The Dream of Yesterday

The dateline is 24 January 1986. The location is Uranus, seventh planet from the Sun. The distance from planet Earth is 1,842,610,000 miles (2,965,400,000 km).

Fifty thousand miles (80,000 km) above the cloud tops of this strange gaseous world, a spacecraft is silently scanning the great globe of Uranus with instruments and cameras. It is NASA's space probe Voyager 2, making its rendezvous, precisely on time, after a nine-year journey through the heart of the solar system that has already taken it past Jupiter, Saturn and their many moons.

Voyager's electronic circuits convert the incoming data into radio pulses and beam them Earthwards from a dish antenna. Even though they travel at the phenomenal rate of the speed of light – 186,000 miles (300,000 km) per second, the signals take 2 hours 44 minutes and 50 seconds to reach their destination, by which time they possess less energy than that of a snowflake falling to the ground!

Large dish antennas such as those at NASA's Deep Space Network at Goldstone, California, collect these signals and feed them into state-of-the-art amplifying and data-processing equipment. At Voyager mission control at NASA's Jet Propulsion Laboratory (JPL) in Pasadena, California, the results slowly come in. Most spectacularly they take the form of color images which can be electronically manipulated to bring out important details in the data.

The JPL team have done it all before, during Voyager's flybys of Jupiter and Saturn. But their expertise and patience is rewarded, as they and astronomers worldwide are astonished at what they see in the first ever close-up images of a planet unknown until the eighteenth century, only just visible from Earth with the naked eye, and featureless even to the largest terrestrial telescopes.

As the Voyager team of planetary scientists, engineers and astronomers disbands, Voyager's instruments are switched off, and it resumes its silent journey into the depths of space. But it is not heading for oblivion – not yet. It is speeding on its way to another rendezvous, this time with Neptune, the near-twin of Uranus, in 1998. On 24 August of that year, should all go well, Voyager should encounter that planet, the first man-made object to do so. Only then will its mission in our solar system be over. Only then will it be free to escape the gravitational clutches of our home star, the Sun, and begin a voyage through interstellar space that could take it many thousands of years hence to other planets of other suns, perhaps even to encounter other beings.

Voyager's brilliant achievement is the culmination of nearly three decades of NASA's exploration of the planets, which began in 1962 with a flyby of Venus. NASA had been formed four years earlier to focus American efforts on the development of hardware and software to further the peaceful exploration of space. But the American effort in space began much farther back in time than that.

'Moony' Goddard

It was in 1898 that the English novelist H.G. Wells published *The War of the Worlds*, one of the first great classics of science fiction. It inspired flights of fancy in many an impressionable young reader, but no-one more so than 16 year old Robert Hutchings Goddard from Worcester, Massachusetts. Unlike others, he turned his flights of fancy into reality.

At much the same time as the Wright brothers were solving the problems of powered flight through the air, Goddard was thinking on a grander scale, of rockets that could soar into the upper reaches of the atmosphere and even venture into space.

Unbeknown to him, a humble Russian schoolteacher in Kaluga, a few hundred miles from Moscow, was thinking along much the same lines and was coming to much the same conclusions. His name was Konstantin Tsiolkovsky. Both men realized that only rockets could be used for space flight and that they would need to be fueled by liquid propellants to give them sufficient power.

Tsiolkovsky, however, was not a practical man. Though he worked out the basic principles of astronautics – the science of traveling in space – and designed suitable rockets, he never built any. By contrast, Goddard was a technical man; he could and did build rockets. By the time he died in 1945 he was to have more than two hundred rocket patents to his credit.

Goddard began experiments in rocketry while he was studying for his doctorate at Clark University in Worcester, Massachusetts, and first attracted widespread attention in 1919, when he published a paper entitled 'A Method of Reaching Extreme Altitudes'. In the paper he outlined his ideas on rocketry and suggested, none too seriously, that as a demonstration a rocket

Above: Thirty years before the first flight of the space shuttle, the British Interplanetary Society were discussing a shuttle-type concept for manned space travel. In this picture by space artist Ralph Smith, a manned space glider has just separated from the first-stage booster and is blasting off into orbit. The booster is about to parachute back to Earth.

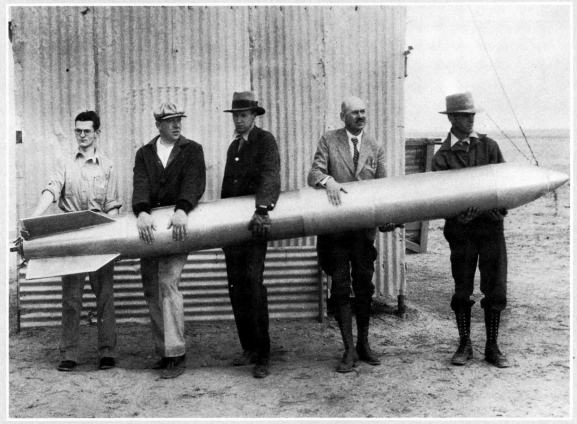

should be flown to the Moon. The public at large, ignoring the scientific content of the paper, latched on to the idea of a Moon flight, which at the time seemed absurd. They more or less dismissed him as just another cranky scientist and nicknamed him the 'Moony'. That taught Goddard a lesson, and from then on he fought shy of publicity.

Goddard soon demonstrated that he was far from being a crank. By 1926 he had designed and built a liquid-propellant rocket. And on 16 March of that year, from a snow-covered field of his Aunt Effies' farm in Auburn, Massachusetts, he fired that rocket into the air. It was the first ever firing of a liquid-propellant rocket, and marked the beginning of modern rocketry. Hitherto all rockets had been fueled by solid propellants, such as gunpowder.

Rocket societies

An interesting link in the story of American space exploration is provided by the American Interplanetary Society (now the American Rocket Society), which came into being in 1930. Formed to promote the cause of space flight, the Society sought to gather and disseminate information on rocketry and discuss seriously the prospects and problems of traveling in space. They wrote to Goddard requesting information about his work, but he tactfully declined. No-one else in the United States was working on rockets, so the Society turned their eyes towards Europe, where rocketry was also beginning to excite groups of enthusiasts.

In the spring of 1931 two founder-members of the Society, husband and wife Edward and Lee Pendray, traveled on vacation to Germany, where they made contact with the German Rocket Society, known as the Verein für Raumschiffahrt (VfR). Formed in

1927, the VfR were already designing and firing their own rockets. A founder-member of the VfR, Willy Ley, took the Pendrays to see rockets being launched at their testing ground, the Raketenflug-platz, in the suburbs of Berlin. The Pendrays returned home, and their enthusiastic report of the VfR launchings initiated the American Interplanetary Society's own experimental program. They test-fired their first rocket, though it didn't leave the ground, in November 1932.

The Pendrays, however, failed to meet another founder-member of the VfR, Rumanian-born Hermann Oberth. Oberth, like Tsiolkovsky and Goddard, was one of the great pioneers of space flight. He summed up his ideas in the classic work *The Rocket into Interplanetary Space* (1923). He was also, directly and indirectly, destined to play a part in the American space program.

Oberth's book enthralled and inspired many people, and changed the course of at least one person's life. That person was Wernher von Braun from Berlin, only 13 when he read the book in 1925. By 1930, with his enthusiasm for space undiminished, von Braun joined the VfR and began to assist Oberth in his rocket experiments. For him rocketry was fast becoming an obsession.

Rockets for the military

In July 1932 the VfR launched a Mirak rocket as a demonstration for the German army rocket research group, headed by Captain Walter Dornberger. Von Braun was present at the demonstration, which took place at the army's proving ground at Kummersdorf, near Berlin. Dornberger was less impressed by the rocket than with von Braun. And in the following October he engaged him to work on liquid-propellant rockets for the army. Von Braun was soon

Above left: Robert Hutchings Goddard poses beside his liquid-propelled rocket just before launching it on 16 March 1926. Fueled by gasoline and liquid oxygen, it burns for 2½ seconds and travels some 185 ft (56 meters).

Above: By 1932 Goddard is building much more advanced rockets. He is seen here (second from right) with colleagues on 19 April, just before a launching. The rocket they hold features gyroscopic control to stabilize the flight.

9

joined by other workers from the Raketenflugplatz and other VfR members. The VfR itself was soon to disband.

Von Braun and the rocket research group he now headed began designing rockets using ethanol (ethyl alcohol) and liquid oxygen as propellants. Their first success came with two flights by their A2 (Assembly 2) rockets in December 1934. A year later an A3 was on the drawing board. It would serve as a test vehicle for a more powerful engine and a guidance and control system.

In 1936, before the A3 was even built, plans were advanced for the A4. This, said Dornberger, was to be a practical weapon, not a research tool. The facilities at Kummersdorf had by now been outgrown, and operations began to transfer to a remote island in the Baltic Sea called Peenemünde.

First the A3 then the A5 blasted off into the skies over Peenemünde, both rockets designed to test components and systems for the A4. Between 1937 and 1941 von Braun's group launched 70 such vehicles, the later ones routinely reaching altitudes of 8 miles (13 km). Technically, the rockets were far in advance of any others in the world, including Goddard's in the United States.

In March 1942 the A4 lifted off the ground for the first time at Peenemünde. But it was not a successful maiden flight. The rocket hardly cleared the cloud tops before crashing into the sea just over half a mile (1 km) away. In the August, a second A4 reached a respectable 7 miles (11 km) altitude but then exploded. On 3 October, however, a third A4 worked perfectly, following its programmed trajectory and landing some 120 miles (190 km) down range.

In retrospect this launching can be seen perhaps as the beginning of the Space Age, for it was the A4 that was the direct ancestor of practically all space launch vehicles, not only in the United States but also in Russia.

A rocket goes to war

The year following the successful debut of the A4, 1943, was a bad one for the German war machine, now locked in combat with the Allies to the West and East. In July of that year German Führer Adolf Hitler suddenly realized he had in the A4 a weapon that could, maybe, turn the military tide in his favor once again. The A4 would be his second vengeance weapon, or Vergeltungswaffe-zwei (V2). His first vengeance weapon, the V1, was a flying bomb powered by a pulse jet, whose rasping sound earned it in Britain the nickname of 'buzz-bomb'.

The V2 weapon program suffered what might have been a fatal setback in August 1943 when Peenemünde was bombed by the British. In the event, Dornberger and von Braun managed to salvage vital plans and equipment, assisted by Oberth, who had joined von Braun two years earlier. Oberth soon left the team, however. Little did he suspect that years later he would rejoin von Braun in the United States and assist in the US space program.

By the end of 1943 mass production of the V2 was underway in Germany, at an underground factory in the Harz Mountains, known as Mittelwerk. In September 1944 V2s were launched against Britain for the first time. The 1200 or so V2s that reached their target, mainly London, over the next few months, caused widespread devastation and killed upwards of 2500 people.

Whereas the Royal Air Force and anti-aircraft batteries had some success in knocking out V1s, they were powerless to combat the V2s, which dropped suddenly and silently out of the sky and impacted on their target at some 3000 mph (5000 km/h).

Thankfully, however, the V2 offensive came too late to markedly influence the course of the war. By April 1945 the German army was in retreat on all fronts. Hitler committed suicide on the last day of that month. Von Braun and the bulk of his V2 team were by this time lodged in Haus Ingeburg, an inn near Oberjoch near the Austrian border. On 2 May advancing American troops were only a few miles away. After initial contacts by his brother Magnus, von Braun and his team of more than 100 surrendered to the Americans. On that day too Berlin fell. Five days later the war in Europe was over.

Over the next few months, as von Braun and the V2 team were interrogated, the Allies began accumulating V2s and V2

Above left: German technicians prepare a V2 for take-off during the closing stages of World War II. The man near the nose is making final adjustments to the flight-control mechanism. The target for this 46-ft (14-meter) long rocket is Britain's capital, London.

components from around Europe, as well as the German technicians who knew how to launch them. In June, Supreme Allied Commander General Dwight Eisenhower sanctioned what were to be the final launchings of V2s in Europe, under the code name Operation Backfire. Backfire culminated in the launching of three V2s from Cuxhaven, Germany, in October 1945. Watching the final launch but from outside the fenced-off launch area and not as one of the official Soviet observers, was Colonel Sergei Korolev. He would reappear some 10 years later as the Soviet Union's 'chief designer of spacecraft', responsible for building the Vostok, Voshkod and Soyuz spacecraft, which since 1961 have carried all Soviet cosmonauts into orbit.

Launchings at White Sands

Though some of his team participated in Operation Backfire, von Braun did not. At the time the launchings were taking place at

Cuxhaven, he was on the other side of the Atlantic, in Fort Bliss, near El Paso, Texas. Piled up in the desert near Las Cruces, New Mexico, were crates of parts enough to build a hundred V2s. Von Braun and some of his team soon moved to the White Sands Proving Ground nearby and began work on assembling V2s from the parts in preparation for launching. By the end of February 1946 the rest of the Peenemünde team had joined von Braun. The US space program was under way.

In March V2s began rolling off the White Sands 'assembly line', and on 16 April the first one was launched, though it had to be destroyed at only a few miles altitude. The second launching, on 10 May, however, was flawless, the rocket climbing on a pillar of flame to over 70 miles (110 km).

Over the next six years, 64 V2s were launched from White Sands, providing the experience in rocketry that laid the foundations for US missile development. The launchings also provided a scientific bonus. The V2 allowed for the first time investigation of the upper atmosphere and indeed the fringes of space. Hitherto, scientists had been limited to low-altitude investigation by balloon-borne instruments. Even before the first launching from White Sands, a V2 Upper Atmosphere Research Panel had been formed (in January 1946) to coordinate scientific operations.

Experiments were contained in the V2 nose cone, now empty of explosives of course, and were also sometimes mounted on the fins. The scientific payloads included such instruments as cosmic-ray telescopes, spectroscopes, cameras, and pressure and temperature sensors, along with their associated electronic circuitry. Often the nose cone was separated with explosives and parachuted back to the ground.

For von Braun and his design team, the White Sands trials were pretty much routine. They were helping to design improved missiles for the services, but nothing on the scale, or with the challenge, of their work on the A4 at the beginning of the decade. Von Braun still had his sights set on loftier things, however. Never since his early days in the VfR had he lost his dreams of traveling in space using rocket technology. In February 1949 he startled some distinguished senior US officers by explaining how rockets could be used to explore space and launch satellites around the Earth.

As if to reinforce his argument, that same month the White Sands team used a two-stage vehicle to reach an altitude of nearly 245 miles (400 km), well beyond the Earth's atmosphere and into space. The two-stage vehicle, known as Bumper, consisted of a V2 with a WAC (Without Any Control) Corporal rocket mounted on it. In all, six launchings of this piggy-back combination, so necessary for space launchings, took place at White Sands. Another link in the US space program was being forged.

With the gradual improvement in the range and sophistication of military missiles, the White Sands facility was fast becoming too small. A new longer range test site was urgently needed. A site was chosen on the, then, remote and virtually uninhabited Cape Canaveral, on Florida's east coast. The Joint Long Range Proving Grounds, as the new facility was called, was established in October 1949. On 24 July of the following year the first launching took place from the Cape. An Army team from White Sands conducted the launch of a two-stage Bumper rocket. It is

interesting to contrast that first launch with the launches that take place at the Cape site today, now known as the Eastern Test Range. Whereas Bumper only rose a modest 10 miles (16 km), today's Atlas-Centaur can launch a spacecraft far beyond the Earth's gravitational influence into the depths of outer space. Launch facilities have also progressed somewhat from the painter's scaffold that served as Bumper's service gantry and the converted tar-paper bathhouse that was its control center!

The super-V2, Redstone

The transfer of launch operations to the Cape coincided with the transfer of the Army missile program from White Sands to the Redstone Arsenal, near Huntsville in Alabama. Von Braun arrived in Huntsville in April 1950 and it was to remain his home for 20 years. The rest of the members of the team soon followed. Once installed, they began work on a 200 mile (300 km) range Redstone missile, essentially a super-V2.

Test flights of the Redstone were carried out at the Cape from August 1953, not always successfully. By mid-1954 the advances being made in rocket technology, not only by von Braun's Army team but also by the Air Force with their Atlas ICBM (Intercontinental Ballistic Missile), began to attract attention in Washington. Would the US soon be in a position to launch satellites into orbit? When consulted, von Braun answered 'yes' emphatically and he outlined how it could be done with existing rockets and missiles within a few years.

The following year (1955) President Eisenhower called for detailed proposals for placing a satellite in orbit as part of the scientific investigations to be carried out during the 1957-58 International Geophysical Year (IGY). Three rival schemes came forward from the three armed services. The Army's (von Braun's) launch vehicle would consist of a modified Redstone rocket with upper solid rocket stages; the project was code named Orbiter. The Air Force suggested their Atlas ICBM, still under development. And the Navy, via the Naval Research Laboratory, lobbied for its Vanguard rocket, based on its Viking high-altitude research rocket.

Von Braun's hopes were dashed when he heard that the Vanguard scheme had been chosen. The reason cited for its selection was that it would be a further development of a scientific research program rather than a converted military missile.

The Army group, however, were not disheartened by this setback, and began working to develop more or less the kind of space launch vehicle they had suggested in Project Orbiter, but this time in the guise of an IRBM (Intermediate Range Ballistic Missile) named Jupiter. That same year von Braun persuaded Oberth to join him. The old team was back together again.

Jupiter in flight

Early in 1956 a new organization was set up at the Redstone Arsenal to develop the Jupiter IRBM. It was named the Army Ballistic Missile Agency (ABMA). The first flight of a Jupiter vehicle took place triumphantly from the Cape in September 1956, when it flew a record-breaking 3335 miles (5335 km) down range. It was a version known as Jupiter C, which had upper solid rocket stages. Experience with this multi-stage vehicle was later to be decisive in inaugurating American space flight.

Vanguard development, meanwhile, was proceeding relatively smoothly with successful test flights in December 1956 and May 1957. It seemed as if the first satellite to orbit the Earth would certainly be American.

But it was not to be. On 4 October came the bombshell. The Soviet news agency Tass announced to the world that the Russians had successfully launched the first satellite (Sputnik 1) into space. Confirmation that a statellite was in orbit came quickly from Britain's newly completed giant dish antenna at Jodrell Bank radio observatory in Cheshire. Among the team celebrating at the Sputnik's Baikonur launch site on the desolate steppes of Russia, was chief spacecraft designer Korolev, uninvited observer of the last V2 launch at Cuxhaven.

In an impromptu speech after the launch, Korolev said that it was 'the realization of a dream nurtured by some of the finest men who ever lived'. He mentioned Tsiolkovsky for one. He could equally well have added Goddard. Both in their time tended to be dismissed as cranks because of their dreams. Time has proved them otherwise. Goddard knew it would. As he said: 'It is difficult to say what is impossible, for the dream of yesterday is the hope of today and the reality of tomorrow'.

Left: Sputnik 1, the little aluminum sphere that put the Russians into space and gave the Americans a nasty jolt when it became the world's first artificial satellite on 4 October 1957. Some 23 inches (58 cm) in diameter, it remained in orbit for 90 days.

Far left: A V2 lifts off the launch pad at the White Sands Proving Ground in New Mexico in the late 1940s. Test firings of V2s brought over from Germany after World War II provided the impetus for US rocket development.

Left: Old pals from the Peenemünde days now working together at the Army Ballistic Missile Agency (ABMA) at Huntsville, Alabama, in 1956. From left to right they are: Ernst Stuhlinger, Hermann Oberth, Wernher von Braun and Eberhard Rees. With them is Commanding Office Major General Holger Toftoy, who was involved in the shipment of captured V2s and the plan (Operation Paperclip) to transfer German scientists, including V2 personnel, to the United States in 1945.

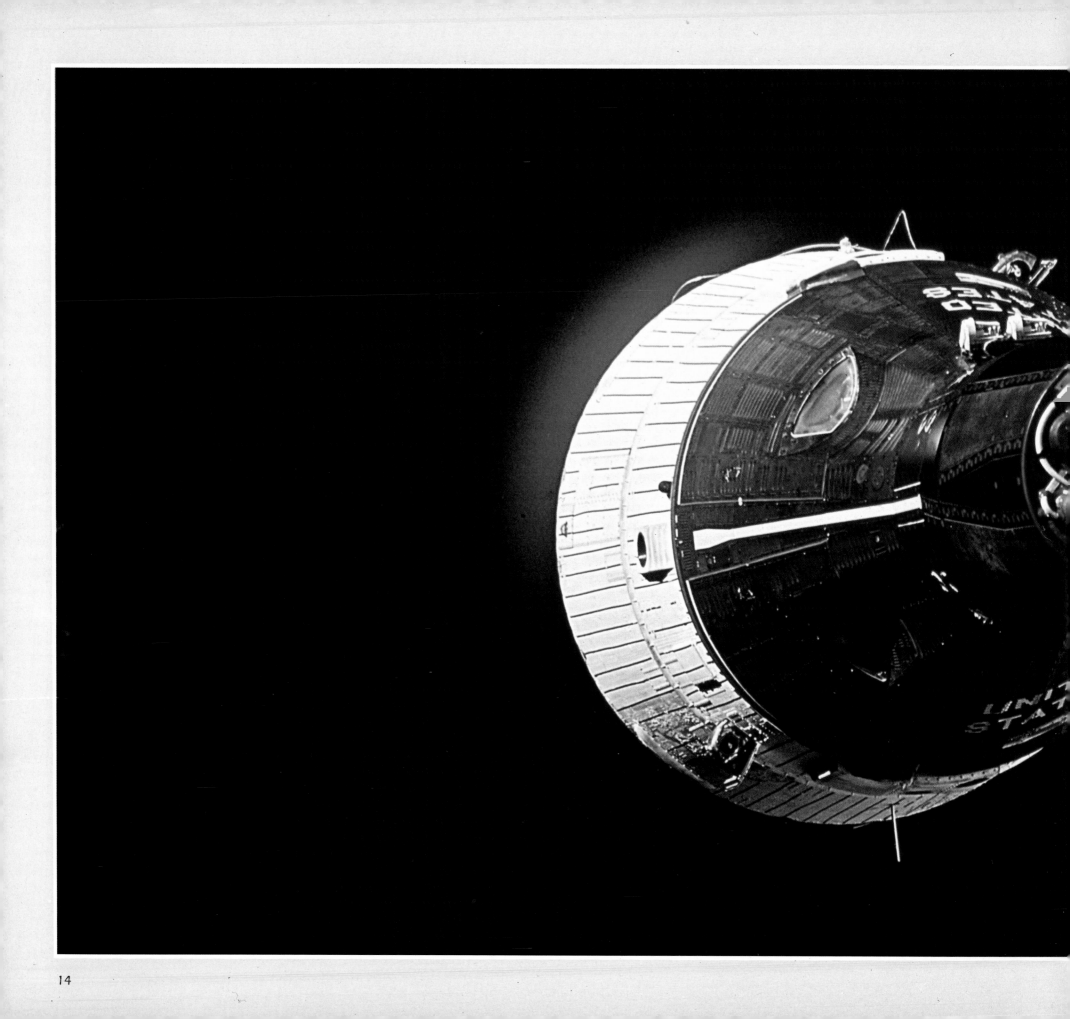

Chapter 2

The Race Is On

As headlines around the world proclaimed the launch of the Space Age with Sputnik 1, certain offices within the US political and military establishments were in a state of turmoil, if not blind panic. That the Russians should have beaten them into space was bad enough, but the size of the satellite they had launched was an unbelievable 184 lb (83 kg). This was five to six times as big as the payload they were trying to launch. This meant, uncomfortably, that the Russians had streaked ahead in rocket and missile design.

As luck would have it, on Sputnik 1's launch day, the new US Secretary of Defense Neil McElroy was visiting the ABMA at Huntsville, where von Braun told him that despite the Navy's continued attempts, the ABMA had the hardware ready for a satellite launch, if required.

On 3 November 1957, however, before any decision to launch could be taken in America, the Russians launched Sputnik 2. As if to emphasize their superior technology, not only did this second satellite weigh a colossal 1120 lb (504 kg), it also carried the world's first live space traveler, a dog named Laika. Within five days von Braun was given the go-ahead to launch his satellite. The time-span required would be dictated, not by the launch vehicle preparation, but by the construction of the satellite to be carried, a task allotted to California Institute of Technology's Jet Propulsion Laboratory.

The decision to duplicate launch attempts increased pressure on the Navy's Vanguard team, struggling at the Cape since 1955 to get into orbit. On 6 December, however, they were finally ready. As Vanguard's engines roared into life, their hopes of achieving an American first in space soared high. Vanguard alas, did not. Less than two seconds into the flight it lost thrust, fell back to the launch pad and ignominiously exploded in a spectacular fireball.

'Goldstone has the bird!'

Spurred on perhaps as much by inter-service rivalry as by the urgent need to restore some vestige of national pride, von Braun's Army team finalized launch preparations for their satellite attempt. On 29 January 1958 the launch rocket was waiting on the pad, ready to go. It was a modified Jupiter C rocket known as Juno 1. The first stage was essentially an updated Redstone. On top were three further stages, consisting respectively of 11, 3 and 1 solid-propellant rockets. The top stage housed the prospective satellite, Explorer 1.

These pages: Some 150 miles (250 km) above the Earth two US spacecraft make the first space rendezvous. They are Gemini 6A and Gemini 7 (right). The date is 15 December 1965. The two craft maneuver within 1 foot (30 cm) of each other.

The 29 January launch had to be postponed because of a high-speed jet stream in the upper atmosphere and generally blustery weather around the Cape. On the 31st, however, conditions were favorable, and at 1.30 pm the countdown began. With only a few minor hitches, the countdown was uneventful. At just after 10.45 pm EST (Eastern Standard Time) Juno 1 lifted off the pad and streaked skywards. The upper stages were successfully ignited in turn. Then came the long wait. No-one would know if the upper stage had achieved orbit until the Goldstone tracking station in California picked up the satellite's signal as it came round the Earth for the first time.

More than one and a half hours later, the time came and went when the Californian tracking station should have reported it. Tension mounted. Then suddenly came the message: 'Goldstone has the bird!' Explorer 1 was in orbit, but had arrived late because it was circling higher than expected. Its orbit was elliptical, with a close point to Earth of 225 miles (360 km) and a far point of 1575 miles (2535 km). Fifty-eight thousand, three hundred and seventy-six times would it travel around the Earth before finally succumbing to the planet's gravity in April 1970.

Compared with the Russian Sputniks, the US satellite was puny. Explorer and the upper stage together weighed less than 31 lb (14 kg). But what the Explorer lacked in size, it more than made up for in sophistication. Its components were highly miniaturized, a situation dictated by the weight limitation imposed by the

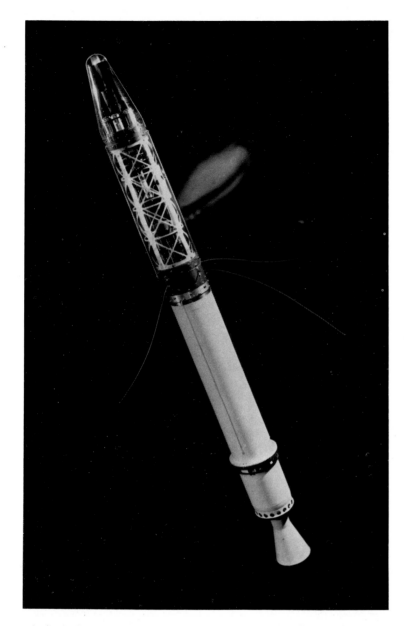

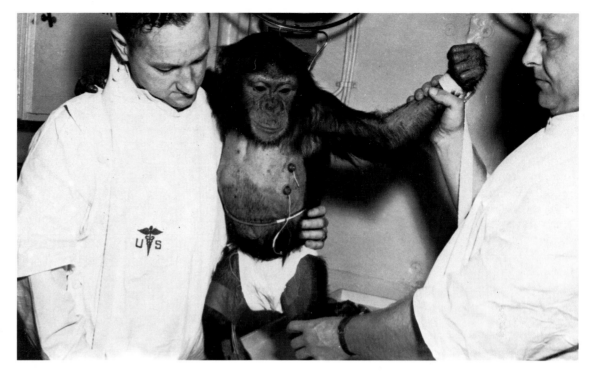

relatively low-powered launch rocket. The Russians, with their powerful rocket boosters, were subject to no such limitations. They had no need to develop sophisticated miniaturized instruments. It was to be their loss, since excellence of instrumentation would prove to be the key element in future space exploration.

Explorer 1 did indeed have excellent instruments, notably a Geiger counter for detecting cosmic rays and other radiation. It reported that a great 'belt' of radiation girdles the Earth. Later satellites confirmed this, finding in fact that there are two major regions of dense radiation. They are now called the Van Allen belts after James Van Allen, who devised Explorer's experiment.

An attempt by the Army team to launch a second satellite with another Juno 1 on 5 March ended in failure. Twelve days later it was the Navy's turn again. This time they were successful, launching a satellite called Vanguard 1 into orbit. The Russians ridiculed this 'grapefruit' of a satellite, and in truth it was scarcely much bigger than one. But, metaphorically speaking, Vanguard 1 had the last laugh. The first satellite powered by solar cells, it

remained active for six years. Not only that, but the prolonged observations of Vanguard in its very stable orbit led to new information about the Earth's shape. Finally, whereas the early Sputniks plummeted back to Earth within a few months, Vanguard 1 is still up there, the oldest satellite in orbit. Not until sometime in the 40th century will the 'grapefruit' satellite eventually return to Earth.

From NACA to NASA

March 1958 was a key month in the history of the US space program for another reason. On the 5th a proposal was submitted to President Eisenhower for the creation of an agency to implement a national space program. The American Rocket Society had called for one a few days after Sputnik 1 was launched (4 October 1957), as had the National Academy of Sciences some weeks later. Now at last this was in prospect.

Who was to run the program, however? Was it to be the military, who had developed the first space rockets, or a civilian organization? The consensus of opinion came down in favor of a civilian agency, especially in view of the fact that the basis for one already existed in the National Advisory Committee for Aeronautics (NACA), a body set up as long ago as 1915.

NACA, which had been created to coordinate aviation research in the US, had by the 1950s become increasingly involved in missile research and was already contemplating moving into space exploration. It had extensive existing facilities around the country, and long-standing relationships with industry and with the military. Above all it had a peaceful, research-orientated image.

So early in April a bill to establish a National Aeronautics and Space Agency, with NACA as its nucleus, to continue aeronautics research and push forward a national space program, went before Congress. During three months of debate on the bill, the structure of the new body was refined, and the 'Agency' became an

Far left, bottom: The Huntsville Times *for 1 February 1958 announces the great news that America is in space at last. The historic first launch the previous day put the Explorer 1 satellite into orbit, where it remained for 12 years.*

Far left: Final preparations are made for the launch of the first US satellite, Explorer 1, on the evening of 31 January 1958. The launching rocket, Juno 1, is a modified Jupiter C, with a fourth stage carrying the satellite.

Above left: Explorer 1, as it would have appeared in orbit, with a rocket stage still attached. The total length is nearly 81 inches (205 cm) and the weight is 31 pounds (14 kg).

Above: Three years to the day after Explorer went into space, chimpanzee Ham is being readied for a suborbital flight in a Mercury capsule. He has already been wired with biosensors to monitor body functions.

'Administration'. Then both houses passed the bill, and President Eisenhower signed the legislation to bring the National Aeronautics and Space Administration into being on 27 July 1958.

NASA set up

When NASA first opened for business on 1 October 1958, it inherited the vast organization that already existed with the NACA, and a work force of some 8000 people. The main facilities comprised Wallops Island Station (now the Wallops Flight Facility) off the coast of Virginia; the Langley Memorial Aeronautical Laboratory (now Langley Research Center) at Hampton, Virginia; the Lewis Flight Propulsion Laboratory (now Lewis Research Center) at Cleveland, Ohio; the Ames Aeronautical Laboratory (now Ames Research Center) at Moffet Field, California; and the High Speed Flight Station (now Dryden Flight Research Center) at Edwards, California.

Over the next few years the structure of NASA as it is today was gradually forged. The various elements of space exploration in existing military research and development establishments were transferred to NASA management and sometimes relocated. First to come under NASA's control in December 1958 was the Jet Propulsion Laboratory, which evolved from facilities where Professor Theodore von Karman of the California Institute of Technology (Caltech) and his students had carried out 'rather odd experiments in the Arroyo Seco north of Pasadena' in the mid-1930s. It had, you recall, built the first US satellite. Caltech operated JPL for NASA, as it still does.

In May 1959 key Vanguard personnel from the Naval Research

Below left: The seven astronauts who will hopefully follow Ham into space are pictured here in their silvery spacesuits. The so-called 'original seven' are from left to right, front row: Walter Schirra, Donald Slayton, John Glenn and Scott Carpenter. Back row: Alan Shepard, Virgil Grissom and Gordon Cooper.

Below right: It was not an American, however, who blazed the human trail into space, but a Russian, cosmonaut Yuri Gagarin, who rocketed into orbit in Vostok 1 on 12 April 1961.

Right: Alan Shepard lifts off the launch pad at Cape Canaveral on 5 May 1961 riding a Mercury-Redstone rocket to make the first American space flight. It is a suborbital arc that takes him nearly 300 miles (500 km) down range.

Laboratory and other space scientists and engineers involved in launching and tracking spacecraft were brought together at a new facility at Greenbelt, Maryland, just outside Washington DC. First called the Beltsville Space Center, it was later renamed the Goddard Space Flight Center in honor of the American rocket pioneer. At the Cape, Vanguard launch personnel formed the nucleus of NASA's first launch team, under the management of the Goddard Center.

Von Braun's Army team at Huntsville, the ABMA, were by this time planning a super booster rocket. The Army decided it was not interested and, in October 1959, the project and von Braun's team were transferred to the NASA space program. The new organization at Huntsville was known as the Marshall Space Flight Center, with von Braun as Director. It was specifically assigned the

task of developing heavy launch vehicles for the burgeoning man-in-space program.

Prominent in this new program were the personnel of the Space Task Group (STG) at Langley. By 1961 this group had expanded to such an extent that new premises were urgently needed. A site just outside Houston was finally selected, and STG personnel began transferring there in the October. The facility was completed within two years. Originally known as the Manned Spacecraft Center, it was later renamed the Johnson Space Center after President Lyndon B. Johnson, who as a Senator played a vital role in the establishment of NASA.

Meanwhile, the Cape was being transformed from scrubland into a major launching center. Launch complexes were springing up the length and breadth of the peninsula. Redstones, Jupiters, Thors, Atlases and Titans – IRBMs and ICBMs – blasted in turn off the launch pads.

By 1962, however, it had become evident that the facilities at the Cape could not support the developing manned space program. After consideration of a variety of possible new launch sites – in California, Texas, on Caribbean islands, and even in Hawaii – Merritt Island, inland from the Cape, was eventually chosen. Work then began to convert this mosquito-infested swamp region into the world's premier spaceport. The new facility, first known as the Launch Operations Center, was later renamed the Kennedy Space Center after the assassinated US President John F. Kennedy. It could not have been better named, for it was Kennedy who committed the nation to the biggest technological challenge in history: to send a human being to another world, the Moon, and bring him back safely to Earth.

The 'original seven'

The official announcement that the United States was to embark on a manned space flight program came on 7 October 1958, just six days after the formation of NASA. After the announcement, NASA's first Administrator Keith Glennan summed up the resolve of the nation when he said: 'Let's get on with it'.

In fact the ball had already started rolling. The essential features of the first phase of the program had been worked out, largely by a team at Langley, later to become the Space Task Group, and a preliminary design for a suitable spacecraft had been selected. By the end of November the program had been christened Project Mercury, after the fleet-footed messenger of the gods in Roman mythology. In February 1959, orders were placed with McDonnell to supply 12 Mercury 'capsules', as the man-carrying spacecraft became known.

At the same time groups of would-be astronauts had been selected by the armed services. Just seven (mainly test pilots) came through the rigorous selection procedures. They were John H. Glenn (Marines); Walter M. Schirra, Alan B. Shepard and M. Scott Carpenter (Navy); L. Gordon Cooper, Virgil I. Grissom and Donald K. Slayton (Air Force). Over the next few years, these prospective astronauts were to undergo intensive training for the largely unknown rigors of spaceflight. It included regular rides in centrifuges, which accelerated them like a rocket would; and

Above left: Minutes after splashdown in the Atlantic, pioneering astronaut Shepard is winched from his floating capsule aboard a recovery helicopter.

Above: A relaxed Shepard pictured with his recovered capsule on the deck of the Navy carrier USS Champlain. *The picture shows just how small the Mercury capsule is.*

Right: This is the Mercury-Atlas launch vehicle that in November 1961 launches chimpanzee Enos into space on a successful two-orbit flight.

Below: An immaculate John Glenn, the ex-fighter pilot selected to be the astronaut on the first manned Mercury craft into orbit. He is later to make a considerable name for himself in politics and run unsuccessfully for President. Note the lapel badge, signifying he is one of the seven original astronauts.

trainers like the Mastif (Multiple Axis Space Test Inertia Facility), which could spin them simultaneously around three axes of rotation, simulating a tumbling spacecraft.

The rocket chosen to launch the Mercury capsule into orbit was the Air Force's Atlas ICBM, the combination being known as the Mercury-Atlas (MA). Redstone rockets were also to be used in the initial flight program, the launch vehicle-capsule combination being known as Mercury-Redstone. These would be used for suborbital flights, that is, those in which the capsule did not go into orbit but arced up through the outer reaches of the atmosphere and returned to Earth in a ballistic trajectory. Both the Atlas and Redstone rockets underwent development to make them reliable enough for humans to fly in. A further type of rocket was also used for testing, known as Little Joe. This was used to carry dummy Mercury capsules in trials.

The flea jump
By the fall of 1959 the various elements of Project Mercury were coming together, and flight-testing began. Flights of the Little Joe rocket, some carrying monkeys, helped test the launch escape system atop the capsule and the tracking network being set up to monitor manned Mercury flights. Mercury-Atlas and Mercury-Redstone flights, two carrying dummy men and one a chimpanzee, verified other spacecraft operational systems.

By 1961 the men and machines seemed to be getting close to a

Right: This plaque on the site of the Mercury launch pad at Cape Kennedy, long since dismantled, commemorates Glenn's historic first flight.

Below: On 20 February 1962 John Glenn squeezes into his Mercury capsule Friendship 7 atop a Mercury-Atlas rocket and prepares for take off. (All the Mercury spacecraft have '7' in their name, for the original seven astronauts). He makes a successful three-orbit flight, though not without drama.

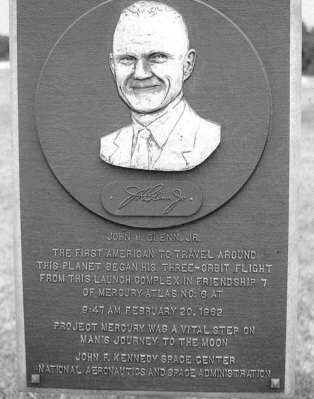

JOHN H. GLENN, JR.
THE FIRST AMERICAN TO TRAVEL AROUND
THIS PLANET BEGAN HIS THREE-ORBIT FLIGHT
FROM THIS LAUNCH COMPLEX IN FRIENDSHIP 7
OF MERCURY-ATLAS NO. 6 AT
9:47 AM FEBRUARY 20, 1962
PROJECT MERCURY WAS A VITAL STEP ON
MAN'S JOURNEY TO THE MOON

JOHN F. KENNEDY SPACE CENTER
NATIONAL AERONAUTICS AND SPACE ADMINISTRATION

Above: In September 1962 President John Kennedy visits the Manned Spacecraft Center, Houston. Appropriately for one who urged the nation to go for the Moon, he is presented with a model of the Apollo command module. On his left is Vice-President Lyndon Johnson, after whom the Manned Spacecraft Center is later renamed.

Right: Gordon Cooper is riding the Mercury-Atlas this time as it accelerates off the launch pad on 15 May 1963. In his capsule Faith 7, he goes on to complete a record 22 orbits, the longest US flight to date.

state of flight readiness. First would come two suborbital flights to man-rate the Mercury capsule. Then would come the real thing – the first manned orbit of the Earth. A chimpanzee named 'Ham' flew in January on a full dress rehearsal for a manned suborbital, and scored a great success, after which the first manned launch was set for late March. But it was not to be.

This proved unfortunate, for on 12 April the US space effort was once again dramatically upstaged by the Soviets. On that day Yuri Gagarin blasted off into orbit to become the world's first astronaut, or cosmonaut as the Russians call their space travelers. The Americans were again seen to be running second in the contest for supremacy in space exploration, now increasingly being seen as a space race between the world's two great powers.

Early in May all was at last ready for a manned suborbital by the Mercury capsule, for which Alan Shepard had been selected. A 2 May launch was planned, but was canceled because of bad weather. A 4 May launch attempt was also scrubbed. Finally, however, on 5 May the go-ahead was given. Shepard, clad in a silvery spacesuit, was maneuvered into the cramped Mercury capsule, a metal cone only 10 feet (3 meters) high and 6.6 feet (2 meters) in diameter at the base.

At 9.34 am local time, after several delays in the countdown, the Mercury-Redstone rocket with Shepard's capsule *Freedom 7* perched on top, thundered from the launch pad. With a heart rate peaking at nearly 140 beats a minute and acceleration forces climbing beyond 6g (six times the pull of gravity), the first American astronaut arced up to a height of over 115 miles (185 km).

After a few minutes experiencing weightlessness at the top of the arc, Shepard began his way down. He now fired the retrorockets to check his descent and the capsule, blunt-end first, plummeted back into the atmosphere, which rapidly slowed the capsule down. It was now that Shephard experienced acceleration forces that peaked at an incredible 11g for a few seconds. But the atmospheric drag had done its job, and soon *Freedom 7* was falling slowly enough to deploy the parachutes that lowered it to a gentle splashdown in the Atlantic. It was still only 9.49. Shepard's historic flight had taken just 15 minutes.

As to be expected, Shepard was showered with congratulations,

spacecraft after Mercury. This was in addition to its prime function of supporting Earth-orbital operations.

In July 1960 the proposed follow-up program to Mercury had even been given a name — Apollo, after the Greek god of light. After Kennedy's speech, Apollo became specifically a Moon-landing project. But it was still ill-defined, with groups at NASA Headquarters (Washington), Marshall Space Flight Center and Langley each favoring different techniques for lunar landing. Both front-runners, however, required a modular craft and needed the crews of the modules to be able to rendezvous and link up (dock) with one another either in Earth or lunar orbit.

It soon became clear that from Mercury to Apollo would be too great a technological jump and that an intermediate stage would be necessary, among other things, to perfect rendezvous and docking techniques. It was with this in mind that NASA announced in December 1961 the development of a new two-man spacecraft. In January 1962 it was given the name Gemini, after the zodiacal constellation of 'the Twins.'

By the time that this announcement was made, the Mercury program had been, slowly, making some headway. In the previous July Virgil Grissom had repeated Shepard's feat with a 15 minute suborbital flight, although this had almost ended in disaster when the Mercury capsule sank just minutes after splashdown. More importantly 'Enos' the chimpanzee had successfully carried out a two-orbit flight in the Mercury-Atlas vehicle in November. This had been a rehearsal for a manned orbital flight, testing all rocket and spacecraft systems as well as tracking and recovery procedures.

Above: Any complacency about the state of the American space program is shattered only a month after Cooper's flight. The Russians launch into orbit their first woman cosmonaut Valentina Tereshkova. She alone makes 48 orbits, far more than all the US astronauts put together!

Right: The comparative sizes of the Mercury (left) and Gemini spacecrafts, the crew capsules of which have a similar bell shape. The Gemini craft, however, has opening hatches above each crew member. Two rear sections carry equipment and a retrorocket pack respectively.

Far right: This dramatic picture shows the interior of the Gemini craft taken through a fish eye lens. There is a little more room for the astronauts than in Mercury, but not much.

receiving a call from President Kennedy on the recovery ship. People throughout the West acclaimed the flight as a milestone. Others, however, were less than enthusiastic. Soviet premier Nikita Khrushchev contemptuously dismissed the flight as 'a flea jump'. And in truth Shepard had only traveled one-eightieth of the distance Gagarin had.

A long way to go

President Kennedy was determined to inject into the space program the same vitality he had injected into his Presidency. On 25 May he delivered a historic speech before Congress that set the seal on future US space exploration.

'Space is open to us now,' he said, 'and our eagerness to share its meaning is not governed by the efforts of others. We go into space because whatever mankind must undertake, free men must fully share.'

'I believe that this nation should commit itself', Kennedy continued, 'to achieving the goal, before this decade is out, of landing a man on the Moon and returning him safely to the Earth. It will not be one man going to the Moon, it will be an entire nation.'

Considering that when Kennedy delivered this speech, no American had even made it into orbit, the call for a Moon landing within nine years seemed to the man in the street a mite optimistic, even unrealistic. But the idea of manned lunar exploration was not new, at least in NASA circles.

Serious discussion about it had begun as early as 1959 at Langley and elsewhere. It was intended that exploration in lunar orbit and maybe on the surface would be a capability of the next generation

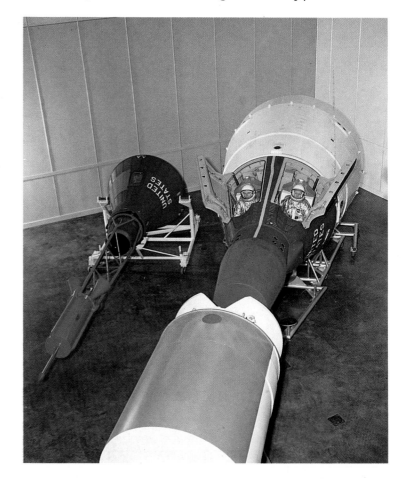

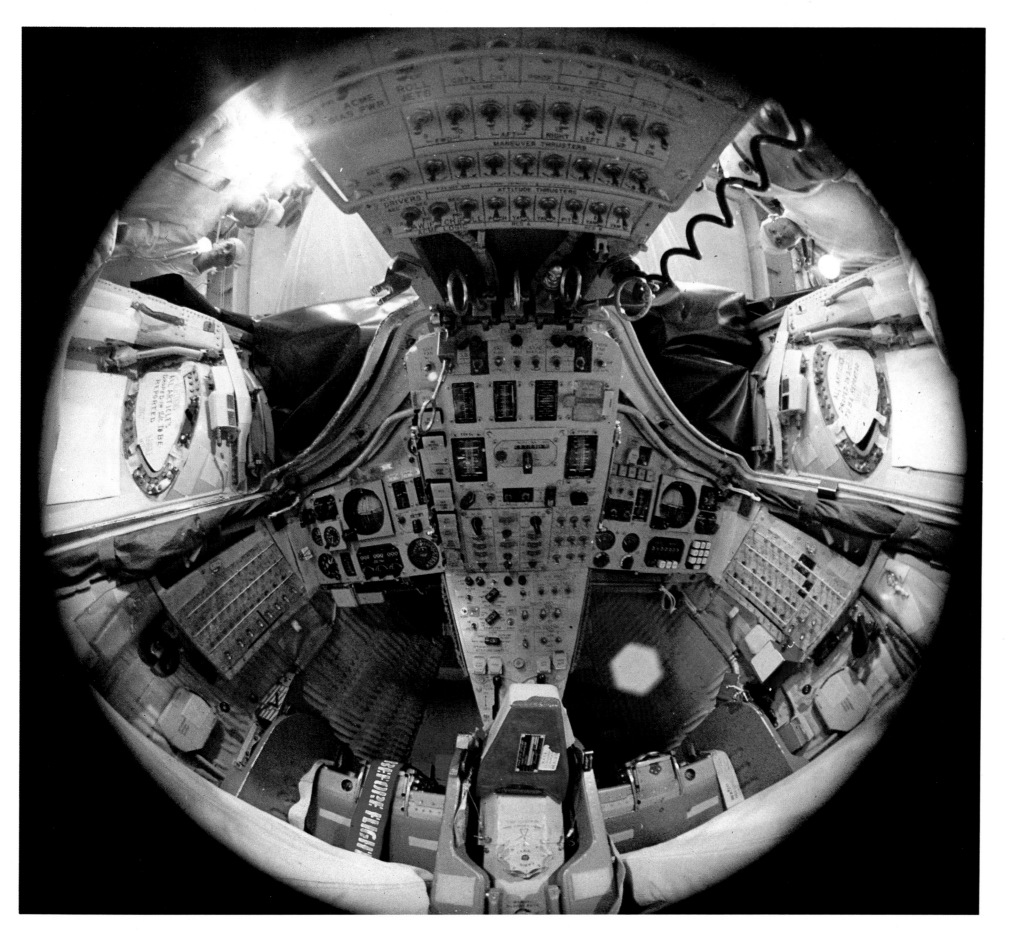

Left: On 23 March 1965 Virgil Grissom and John Young fly into orbit in Gemini 3 to make the first manned Gemini flight. It is the first of 10 flights that enable the United States to nose ahead in the space race for the first time.

Opposite: On 3 June 1965 Edward White emerges from Gemini 4 and makes the first American walk in space. He remains outside his craft for more than 20 minutes.

Below: Against a cloudy backdrop, White fires the gas gun in his right hand to maneuver on his pioneering spacewalk. Note the gold-covered umbilical/tether, which connects his suit with the life-support system of the Gemini 4 craft.

'We're underway'

The day all America had been waiting for finally arrived. After many nerve-racking delays, 20 February 1962 became the date that the United States would attempt to launch a man into orbit. The man chosen to be hero, or martyr, was John Herschel Glenn Jr. Originally from Cambridge, Ohio, Glenn had served as a fighter pilot in both World War II and Korea, and had latterly flown as a test pilot.

At 6 am on the 20th, Glenn eased himself into the couch of the cramped Mercury capsule, *Friendship 7*, perched atop the modified Atlas ICBM on launch pad 14 at Cape Canaveral. The countdown was going well. Outside the Mercury launch area and along the beaches of the Banana River and the Canaveral sea shore, people were gathering in their tens of thousands. Many had been camping out for days, suffering the agony of delays with the launch team itself. This time, however, they were not to be disappointed.

As the countdown reached zero at 9.47 am, Glenn felt the jolt as the rocket ignited beneath him. 'The clock is operating,' reported Glenn, 'we're underway.' Burning a mixture of kerosene and liquid oxygen at a prodigious rate, the Atlas blasted skywards. Glenn felt the g-forces build up to nearly 8g. Five minutes after the lift-off the engines cut out, the g-forces evaporated and he was weightless.

Glenn, traveling at over 17,500 mph (28,000 km/h), was in orbit. At first he was cleared for seven orbits, but this was reduced to three when problems began to arise. After one orbit, which took

an hour and a half, he reported that the rocket thrusters of the automatic attitude control system were malfunctioning, and he had to take manual control. Later, ground control received telemetry information from the spacecraft which showed that the landing bag had deployed. This was a bag that was designed to drop down to cushion impact just before landing. Beneath it was the heat shield. If the bag had deployed in orbit as indicated, then Glenn was as good as dead. The bag and heat shield would be torn away when the capsule re-entered the atmosphere, and there would be nothing to protect Glenn and his capsule from the searing temperatures of up to 1600°C (2900°F) created by aerodynamic drag.

'A real fireball'

Ground control didn't tell Glenn what was worrying them about the heat shield until minutes before re-entry. Shortly afterwards, as he hit the air, he began to see chunks of debris that he thought came from his heat shield flying past the capsule window. The cabin temperature began to rise. After all the work and preparation, was Glenn to become a martyr after all? The ground controllers back at the Cape were out of contact with Glenn now because of the communications blackout that always occurs on re-entry. Agonizing minutes passed by, and then Glenn's voice crackled through the static: 'My condition is good, but that was a real fireball, boy!' The telemetry information about the landing bag had been wrong.

The remaining stages of the flight, the parachute deployment, splashdown and recovery – all went off as planned. The astronaut had been aloft for a total of 4 hours and 55 minutes. In the days that followed, Glenn was feted with tickertape parades and motorcades in Washington, where he addressed Congress; and New York, where he addressed the United Nations.

Over the next two years or so three more American astronauts followed Glenn into orbit. Scott Carpenter (May 1962) made another 3-orbit flight, also experiencing its share of problems, with an overheating suit and temperamental retrorockets. By contrast Walter Schirra's flight (October 1962), though twice as long, was virtually trouble-free. This made the Mercury team confident enough to aim for at least a day in space for its final mission. This would be a worthy achievement for the first phase of the American manned space flight program. On 15 May 1963 this ambition was achieved when Gordon Cooper spent not just 24, but 34 hours in space, orbiting the Earth no fewer than 22 times.

It was with this success that Project Mercury drew to a close. The missions proved that human beings could be launched into orbit and recovered safely, and that they could function in weightless conditions as pilot, engineer and experimenter – at least for 34 hours at a time – with no deterioration of their bodily condition.

So far, so good. But how did the Americans stand at this point in time in the space race? Still unfortunately in second place, by quite a long way. It is interesting to compare the achievements of Mercury with those of Russia's first manned spacecraft program, Vostok. Both programs ended in 1963. Mercury had put four men into orbit for a total of just over two days. Vostok had put five men and one woman into orbit for a total of over 15 days! The Russian woman cosmonaut, Valentina Tereshkova, alone had spent more time in space than all the Mercury astronauts put together.

Two by two

Away from the public eye the Gemini program had been progressing satisfactorily, the initial research had been done and the spacecraft were being built. The Gemini craft was somewhat bigger than the Mercury to accommodate an extra crew member and extra equipment.

Unlike Mercury, Gemini was a modular design. The astronauts

occupied the crew module, or re-entry capsule. In appearance it looked much like Mercury. Behind the crew capsule was an equipment module, and behind that a retro-module that housed the retrorockets. Before Gemini returned to Earth, the equipment and retro-modules would be jettisoned. Weighing nearly double the Mercury capsule, Gemini would require a more powerful launch rocket and this would be found in the Air Force Titan ICBM.

Another major difference between the two spacecraft was that Gemini would be fitted with opening hatches above the astronauts. This would allow, among other things, for the astronauts to leave their craft in space to perform EVA (extra-vehicular activity), or spacewalks. This EVA capability also required major changes in the capsule's life-support equipment. The astronauts' spacesuits would also have to be extensively modified, to afford greater protection from the hostile space environment the men would be directly exposed to for the first time. All these, and many other innovations, would be needed if the Gemini missions were to fulfill their objectives. Essentially, these objectives were to subject men and supporting equipment to long-duration flights; to practice and perfect in-orbit techniques for maneuvering, rendezvousing and docking with other craft; to perform EVA and to assess the prolonged effects of weightlessness on the human body. These ambitious objectives would have to be achieved, and achieved well, for a successful Moon landing to be accomplished within the decade.

Upstaged again

In April 1964 the first flight took place of the Gemini-Titan combination, unmanned. It was a complete success, and provided valuable training for the launch and tracking teams. A manned flight was provisionally scheduled for November. But, as usual, dates slipped and not until 23 March 1965 was the first manned Gemini (Gemini 3) ready to lift off.

But it would not be, as NASA had planned, the first space flight by a multiple crew. The Russians had again beaten them to it, twice. In the previous October they had launched their Voshkod 1 craft with not two, but three people on board. And only five days before Gemini's maiden flight, Voshkod 2 carried a crew of two into orbit, one of whom (Alexei Leonov) added insult to injury by performing the world's first spacewalk! The Russians still appeared to be far ahead. What would they do next?

So it was with the success of the US space program seemingly riding with them, that the Gemini 3 astronauts blasted off the launch pad at 9.24 am on 23 March 1965. Within the capsule were Virgil Grissom, who was to die tragically two years later, and John Young, destined to become commander on the first space shuttle flight. During the five-hour flight the crew changed their craft's orbit three times, the first occasion any manned craft had maneuvered in orbit.

The whole flight went as planned, proving the space worthiness of the new design. The only unscheduled activity occurred when Young nonchalantly produced a corned beef sandwich he had smuggled on board to supplement the nutritious but tasteless

Above: Gemini 9 abandons plans to dock with the Agena target vehicle when a shroud covering the docking mechanism fails to separate. Astronauts Thomas Stafford and Eugene Cernan dub it the 'angry alligator'.

Left: On the Gemini 12 mission in November 1966 spacewalker Edwin Aldrin works at a 'busy box' on an Agena target vehicle. A steady work pace and frequent rests allow him to perform better than any spacewalker so far.

regulation space food! Ground control were less than ecstatic, visualizing what effect weightless crumbs might have on Gemini's delicate control systems.

At the conclusion of the mission, splashdown was some 50 miles (80 km) off-target, but at least the spacecraft kept afloat, unlike Grissom's first one! Tempting fate, he had earlier unofficially named it the 'Molly Brown' after the 'unsinkable' heroine of a popular Broadway musical.

'Worth a million dollars'

On 3 June the second manned Gemini (Gemini 4) took to the skies, aiming for a four-day mission, with James McDivitt and Edward White as crew. White would also lose his life with Grissom two years later. On the spacecraft's third orbit, White opened the hatch and, at the end of a 25 foot (7.5 meter) umbilical-tether, drifted slowly out into space. The view, he said, 'must be worth a million dollars!'

Maneuvering by means of a pressurized gas gun, White tumbled head over heels, reveling in the weightless gyrations only one man before (Leonov) had experienced. His 17,500 mph (28,000 km/h) gymnastics came to an end after 22 minutes. He was exhausted by the time he returned to his seat and closed the hatch. Was fatigue, the NASA experts on the ground wondered, to be a problem with spacewalking? If so, it might jeopardize future space exploration.

Gemini 4's splashdown was, like Gemini 3's, some 50 miles (80 km) off target. The splashdown of Gemini 5, on 29 August 1965, was nearly twice as far off. But this was a minor blemish on an otherwise near-perfect 8-day mission. The flight also achieved a hardware first in carrying fuel cells to provide electrical power. They have been standard equipment on every US manned spacecraft since, providing as a bonus drinking water for the astronauts.

Gemini 7 was the next to blast off into space; its mission designed as an endurance trial to simulate the time it would take to get to the Moon and back. It remained aloft for two weeks in December. It was during this time that Gemini 6A, postponed from October, was launched to rendezvous with it. The two Gemini crews maneuvered to within 1 foot (30 cm) of one another and flew in this close formation for six hours.

The remaining Gemini flights concentrated on rendezvous and docking maneuvers with unmanned Agena target vehicles and included more extensive spacewalking. The Gemini 8 mission in March 1966 saw the first attempt at docking in orbit with an Agena. Docking was achieved but had to be terminated minutes later when Gemini's maneuvering thrusters started firing uncontrollably, setting the Gemini-Agena combination gyrating wildly. Gemini 9 had no luck either with docking, finding that a protective shroud had failed to separate from the target vehicle, giving it the appearance, the crew said, of an open-jawed 'angry alligator'.

On the remaining Gemini flights, however, Agena dockings were successful. By the time the Gemini program drew to a close, with Gemini 12 in November 1966, the astronauts had perfected rendezvous and docking maneuvers. It seemed also that spacewalking fatigue could be combated too. Gemini 12 astronaut Edwin Aldrin was able to spacewalk for over 2 hours, beating

fatigue, spacesuit overheating and foggy visors by taking regular rests.

By the conclusion of the Gemini missions and to everyone's relief, the accuracy of splashdown had improved enormously. The most accurate was that of Gemini 9A, a mere 600 yards (600 meters) from the recovery ship. Other bonuses from the Gemini program included the carrying out of some 40 experiments, notably in space medicine and photography. The astronauts' photographs of the Earth's surface showed up startling detail and demonstrated what benefits comprehensive satellite photography could bring.

The ten manned flights of the Gemini program had achieved all their objectives and more, and they had thrust the United States to the forefront in the space race. For, strangely, there had been no Russian manned launchings at all since the beginning of the Gemini flights. The Russians, like the Americans, were developing a new spacecraft. Both nations were poised at the end of 1966 for the next leap forward – the United States was sprinting for the Moon, perhaps Russia was too. But despite the technological achievements, 1967 was going to be a bad year for both countries.

Above: Through the open hatch of Gemini 12, Edwin Aldrin took this picture showing the docked Agena. The land mass on the horizon is Mexico. With this mission, the Gemini program comes to an end, preparing the way for the great leap of the Apollo program.

Right: High over New Mexico is Gemini 12's Agena target vehicle. A tether has already been fixed to it from the Gemini. Soon the tether will be pulled taut, and attempts will be made to rotate the tethered combination to simulate gravity.

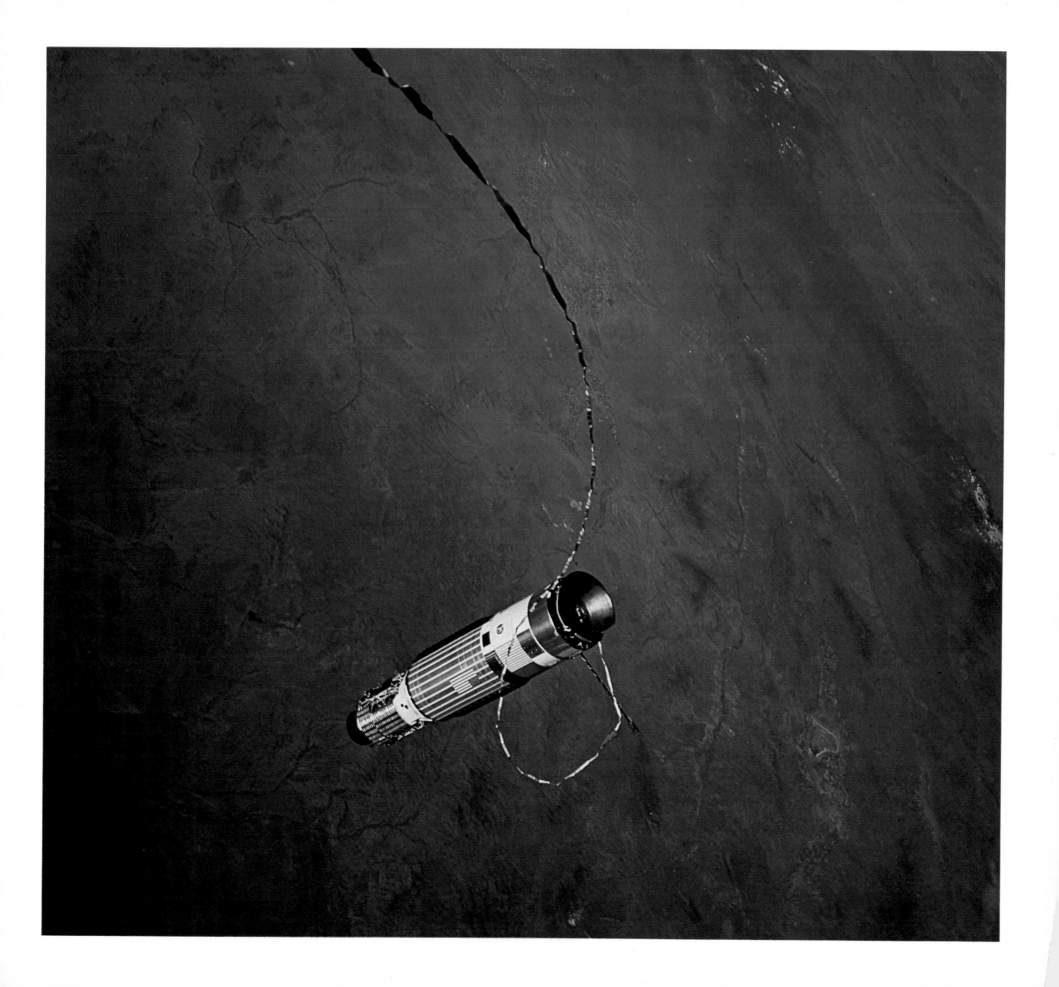

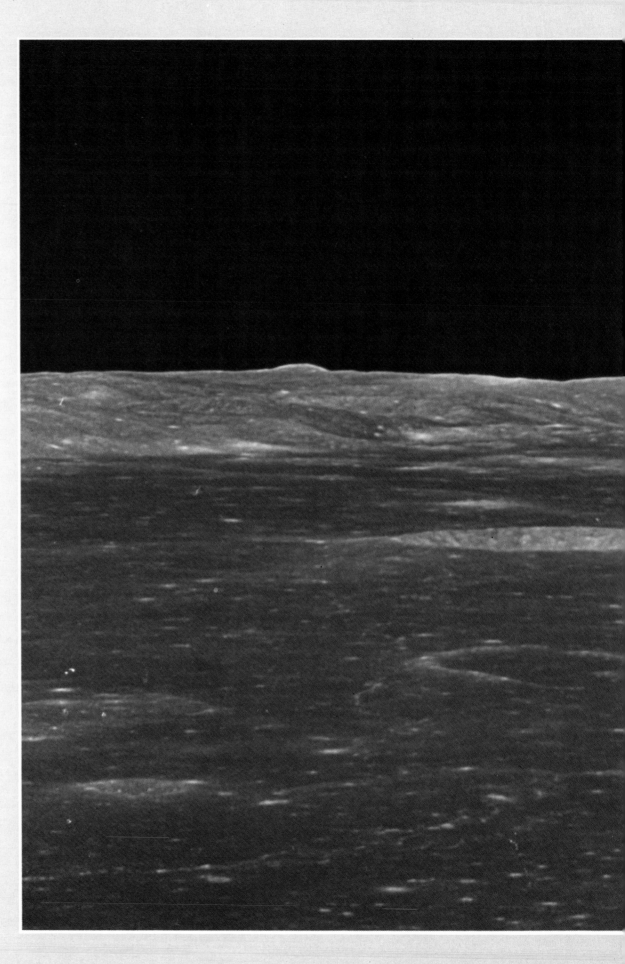

Chapter 3

The Journey to Tranquillity

As mentioned before, Apollo became a Moon-landing program after President Kennedy's famous speech before Congress in May 1961. But the method of reaching the Moon had yet to be decided. Around NASA's various establishments, arguments for and against alternative methods of achieving a lunar landing raged back and forth for more than a year.

There were three major contenders: a direct flight, an Earth-orbital rendezvous or a lunar-orbital rendezvous. The direct mode envisaged a colossal booster (tentatively named Nova) that would carry astronauts directly to the Moon and back. The drawback was that Nova would need to have a thrust in excess of 11 million pounds (5 million kg). The Atlas booster currently being used for Mercury missions developed less than one-thirtieth of this power! Given the limited time available, direct flight was a non-starter.

Earth-orbital rendezvous (EOR) required two launchings of spacecraft into Earth orbit with somewhat smaller rockets than Nova. A mooncraft would be put together there and then blasted off to the Moon. This seemed to have much to recommend it, though it required precision rendezvous and docking techniques and probably the transfer of fuel in orbit, as well as two expensive boosters.

The third method, lunar-orbital rendezvous (LOR), required just a single high-powered launch rocket and a modular spacecraft. This would be placed in orbit around the Moon and then drop a small ferry craft down to the surface. On return, the ferry would rendezvous and dock with the mother ship. By the end of 1962 LOR won out, and detailed design work on it began.

Within a year the Apollo spacecraft design was all but finalized. The three-man crew would occupy a conical command module about 11.5 feet (3.5 meters) tall and 13 feet (4 meters) in diameter at the base, where the heat shield would be located. For most of the mission it would be docked with a 14 foot (4.3 meter) long service module, carrying a propulsion motor, fuel cells and other equipment. Together they formed the CSM (Command and Service Module), the Apollo mother ship.

These pages: Ablaze with light and color, distant planet Earth rises above the drab, barren lunar horizon. Through stunning photographs like this, we can share the experiences of the twenty-seven American astronauts who were privileged to witness the phenomenon live during the Apollo missions.

The ferry ship which would carry two astronauts down to the Moon's surface was called the lunar module (LM). It consisted of upper and lower stages, each with its own engine. It would use the lower-stage, descent engine for retrobraking as it descended to land. The upper stage would use the lower stage as a launch pad when the time came to return to lunar orbit to rendezvous with the mother ship. The total Earth weight of the whole Apollo craft, CSM plus LM, was some 45 tons (45 tonnes).

Providing the muscle

The booster rocket needed for a lunar landing was not required to be as powerful as the proposed Nova, but was still going to have to be a monster. Von Braun's team at the Marshall Space Flight Center was given the go-ahead to develop the Moon rocket early in 1963. It was to be named the Saturn V.

Von Braun had begun developing the Saturn series of heavy launching rockets even before NASA was formed. The first fruits of this labor had been the Saturn I, an ungainly looking vehicle with a cluster of eight propellant tanks around a center core, which had made a successful maiden flight in October 1961. Though derogatively referred to as 'clusters last stand', the Saturn I packed a 1.5 million pound (680,000 kg) thrust.

By 1966 it had developed into the Saturn IB. This had similar, but slightly more powerful, first-stage engines, burning kerosene and liquid oxygen, and a single-engine upper stage burning liquid hydrogen and liquid oxygen. This liquid hydrogen engine marked a new departure in launch rocket design, and the same combination is now used in the space shuttle.

The Saturn IB became a vehicle for testing components for the Saturn V and for launching unmanned Apollo spacecraft for flight tests. It was also man-rated and was subsequently used for launching Apollo, Skylab and Apollo-Soyuz Test Project (ASTP) astronauts.

The Saturn IB stood 140 feet (43 meters) high on the launch pad, and was considerably bigger than its predecessors. But it was dwarfed by the Saturn V, which stood an incredible 365 feet (111 meters) tall. The first stage of Saturn V alone was as big as the Saturn IB. It housed five engines burning kerosene and liquid oxygen. The second and third stages, with five and one engine respectively, burned high-energy liquid hydrogen/liquid oxygen propellants.

On top of this enormously powerful rocket stack, the Apollo

Far left: A complete Apollo stack emerges from the Vehicle Assembly Building (VAB) at the Kennedy Space Center en route for launch pad 39A. The 526 foot (160 meter) tall VAB, capable of holding four Saturn Vs at once, will later be used to process the hardware for the next generation launch vehicle, the space shuttle.

Above: A second-stage rocket for the Saturn V launch vehicle arriving at Kennedy. It has been transported by barge via the Intercoastal Waterway from NASA's Mississippi Test Facility.

Inset above: Other parts of the Apollo hardware are flown in to Kennedy by a special aircraft known as the Super Guppy.

Left: The command and service module (CSM) of Apollo 7 being mated with the lunar module adapter at Kennedy before installation atop the Saturn V booster. Apollo 7 will be the first manned flight test of Saturn V/Apollo.

Above: The solid rockets of the Apollo launch escape tower belch smoke and flame as they carry away a dummy command module during testing of the emergency launch escape system. The tower stands at the top of the Apollo stack on lift-off.

Right: On a simulated lunar landing site at Kennedy, Apollo astronauts Harrison Schmitt and Eugene Cernan rehearse techniques for their forthcoming mission. They will find working on the Moon somewhat different because of the low gravity, which is only one-sixth that of Earth.

spacecraft perched, command module uppermost with an escape tower attached. This could lift the module with its astronauts clear in an emergency. The total lift-off weight was a fantastic 3200 tons (2900 tonnes). Yet Saturn V's engines, developing nearly 3800 tons (3500 tonnes) of thrust, could lift the prodigious load with ease.

A NASA press release graphically described the Saturn V design. 'The vehicle would be longer than a football field, have a base diameter greater than the combined widths of three tractor-trailer rigs, weigh more than a Navy cruiser and develop more power than a string of Volkswagens from New York to Seattle.'

The birth of Complex 39

In conception, the Apollo project was gargantuan, and demanded for its success such gigantic hardware as the Saturn V. And, naturally, everything else connected with this behemoth of a rocket tended to be gigantic too.

The Saturn V was going to be assembled in a hanger so that every one of its over three million working parts could be checked before it reached the launch pad. So a mighty large hanger would be required. As well as this, a huge vehicle would also be needed to provide the transport from the hanger to the pad. The whole launch site was located on Merritt Island, inland from the launch complexes that lined Cape Canaveral. Designated Complex 39, it was to become the focal point of what is now the world's busiest spaceport, the Kennedy Space Center.

Construction of the Saturn V 'hanger', called the Vehicle Assembly Building (VAB), got underway in 1963 and was completed two years later. More than 50,000 tons (45,000 tonnes) of steel were used to build the 526 foot (160 meter) tall VAB, which was designed with four high bays, each capable of handling a Saturn V. Transport for the Saturn Vs was provided by an eight-crawler tracked transporter, the like of which had never been seen before. Two were built, and they are now used for transporting the assembled space shuttle stack. They are still the world's biggest land vehicles.

'Fire in the cockpit!'

During 1966, while Gemini was accumulating vital data for the forthcoming Apollo missions, the launch teams at the Cape started testing the Saturn IB, mated with dummy Apollo spacecraft. All the test flights were successful, and it was decided to proceed with a manned launch in February 1967. The crew would be old hands Virgil Grissom and Edward White, partnered with rookie Roger Chaffee. At about 1 pm on 27 January the crew entered their Apollo craft atop the Saturn IB assembled on launch Complex 34 at the Cape. They were to take part in a simulated countdown for the following month's flight. Fully suited up, they were breathing pure oxygen as they would in real flight. The cabin too was pressurized with pure oxygen.

The countdown did not proceed smoothly because of communications problems between the Apollo spacecraft and ground control and other irritating bugs in the systems. One by one, however, these difficulties were overcome. At 6.31 pm, the countdown was about to resume after one of the many hold-ups. Then suddenly a voice from the spacecraft came over the intercom. 'We've got a fire in the cockpit!' Within seconds the spacecraft was

engulfed in sheets of flame, feeding voraciously on the pure oxygen atmosphere.

Strapped in their couches, the crew had no time even to begin to unbolt the hatch, for the cabin had become an inferno. 'We're burning up' was the last agonized cry from one of the astronauts, and then the cabin ruptured under the pressure that had built up inside. Flames and thick black smoke billowed out, filling the 'white room' that gave access to the spacecraft. The danger worsened, for the escape rocket on top of the cabin might be set off, and if that happened the whole service structure and launch rocket could ignite. Despite the impending danger, some of the technicians ran to fight the fire and open the hatch. But they were too late. No-one could have survived such a conflagration. Grissom, White and Chaffee were dead.

Rising from the ashes

As the nation mourned the horrific death of three of its heroes, a review board sought to find the cause of the fire. They eventually concluded that it must have been triggered off by a minor electrical short in the equipment or wiring. Their far-reaching final report was also critical of aspects of the spacecraft's design, test procedures, and even NASA management practices.

The review board presented its report in April 1967. While NASA

was studying it and considering what to do, news came in on the 23rd that Russia, too, had suffered a tragic loss. On the first test flight of their new spacecraft, Soyuz 1, the parachutes used for braking the descent of the re-entry capsule became entangled; the capsule hit the ground at full speed, killing cosmonaut Vladimir Komarov. Komarov became the first in-flight space casualty.

After these deaths neither the United States nor Russia would attempt another manned mission for a year and a half.

NASA, as a result of the review board's findings, ordered extensive modifications to the Apollo command module. The hatch was to be redesigned for quick opening. Exposed wiring and piping were to be re-routed and covered, and a flame-proof material called beta cloth was to replace nylon in the astronaut's spacesuits. This, of course, meant more delay. With the decade rapidly drawing to a close, could NASA meet President Kennedy's deadline for a Moon landing?

Fortunately, the delay with the command module did not affect the development of the Saturn V. In November 1967, the first Saturn V lifted from Pad A of the newly completed Complex 39. With all stages 'live', which was unusual for a first flight test, the massive rocket performed perfectly. For the next six years, the thunderous roar and earth-shaking vibrations of a Saturn V lift-off would be a regular occurrence at the Cape.

Above: The Apollo 7 astronauts move in to practice docking maneuvers with the upper stage of their launch rocket during the first manned Apollo flight in October 1968. The white disc inside the rocket stage is a simulated docking target similar to that which will be carried on the lunar module on later missions. The maneuvers take place in Earth orbit over the Mississippi Gulf coast.

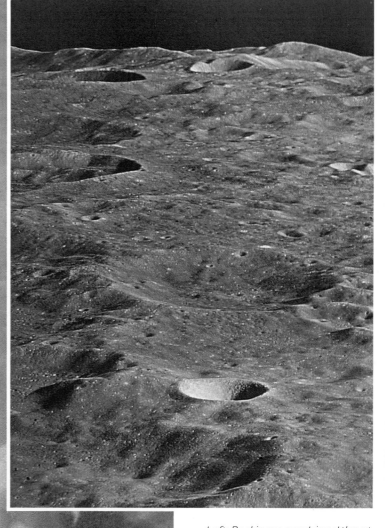

Left: Packing a combined thrust of over 7½ million pounds (3.5 million kg), the five first-stage rocket engines of the Saturn V lift the Apollo 8 spacecraft off the launch pad on 21 December 1968. Of the enormous 365 foot (111 meter) vehicle, only a small cone a few feet across will return six days later after the first circumnavigation of the Moon.

Above: From lunar orbit Apollo 8 sends back this picture of the far side of the Moon. The lunar far side consists mainly of heavily cratered rugged highlands. Unlike the side we can see from Earth, it does not have any large flat mare (sea) regions.

Right: During the Apollo 9 mission in Earth orbit in March 1969 the CSM (code named Gumdrop) is pictured through a window of the lunar module (Spider). The docking mechanism is visible in Gumdrop's nose.

Below: In a full dress rehearsal for a lunar landing in May 1969, the Apollo 10 lunar module Snoopy moves in to dock with the CSM Charlie Brown over the far side of the Moon. The mission verifies all rendezvous and docking maneuvers in lunar orbit. The stage is set for the first lunar landing.

This flight, which also tested an unmanned Apollo CSM, was the fourth in the Apollo/Saturn series and was designated Apollo 4. In 1968 two more proving flights took place which tested the lunar module in orbit and man-rated the Saturn V. In October of that year the time came for the first manned flight of an Apollo CSM, Apollo 7. Blasted aloft on a Saturn IB, the CSM performed perfectly, rather better in fact than the astronauts, who went down with head colds within the first few days of their 11 day mission!

Even as Apollo 7 splashed down, the final preparations were being made for the next mission, Apollo 8. Originally this was to be a thorough testing of the complete Apollo spacecraft, the CSM and the LM, in Earth orbit. But development problems with the LM precluded a scheduled December launch, and so the intended follow-up mission was brought forward in its place. This was to attempt a circumnavigation of the Moon. For the first time in history human beings would leave the protective cradle of Earth and come under the influence of the gravity of another world.

The astronauts chosen for this epic mission were Frank Borman and James Lovell, veterans of the Gemini program, together with rookie William Anders. They would be flying just an Apollo CSM, without the troublesome LM, atop a Saturn V. All would be well if the Saturn V, not yet flown on a manned mission, worked perfectly; if the CSM worked perfectly; if the complex calculations of outgoing and return trajectories proved accurate. But if things did start going wrong, the crew might be doomed to drift for ever in interplanetary space.

Rings around the Moon

There was no feeling of impending disaster, however, only elation, when Apollo 8 lifted off from Complex 39, within seconds of its scheduled time, 7.51 am EST on 21 December. In less than 12 minutes Saturn's three stages had heaved the astronauts into a 'parking' orbit some 120 miles (190 km) high. There the crew and mission control at Houston, which took over control of the flight immediately after lift-off, made a final check of the systems of the CSM and the third rocket stage, which was still attached. They also made the final trajectory computations for the translunar flight.

Nearly 3 hours into the flight, the third-stage engine roared into life again and increased their speed from less than 17,500 mph (28,000 km/h) to nearly 24,400 mph (39,000 km/h); then the third stage separated. Three human beings in a fragile 'tin can' were on their way to the Moon, some 225,000 miles (360,000 km) away, traveling at a speed no man had ever reached before. There was no turning back now.

By the afternoon of the 23rd, the pull of Earth's gravity had slowed Apollo 8 down to less than 2300 mph (8600 km/h). Then the spacecraft started to speed up again as it came increasingly under the influence of lunar gravity. On the morning of Christmas Eve, it was only a few hundred miles from the Moon. Soon a critical maneuver would have to be made. The service module's engine would fire to slow down the craft so that it could be captured by lunar gravity and go into orbit. Since this had to happen while Apollo 8 was traveling behind the Moon, mission control would

Far left: On 16 July 1969 the Apollo 11 astronauts, led by Armstrong, leave for the launch pad. They are fully suited up and carry ventilators to supply their suits with air until they can connect with the on-board life-support system.

Left: This dramatic launch tower photograph captures the historic moment when Apollo 11 leaves planet Earth in an attempt to achieve the seemingly impossible, a manned lunar landing. It is 9.32 Eastern Daylight Time on 16 July 1969.

Right: As Apollo 11 speeds towards the Moon, the astronauts see their home planet recede into the distance. They can make out the continents quite clearly, silhouetted against the azure blue of the oceans.

have no way of knowing if the 'burn' had been successful until Apollo emerged from its lunar eclipse.

In the event the burn was good and Apollo 8 began orbiting around Earth's nearest neighbor in space. On Christmas Eve the crew made two live telecasts from lunar orbit. TV viewers back on Earth watched spellbound as the barren and cratered surface of the Moon unfurled beneath the camera lens, and Earth rose over the lunar horizon. During one broadcast the crew read movingly from the Book of Genesis.

Early on Christmas morning nail-biting tension returned as the crew prepared to fire Apollo's engine to cut them loose from lunar gravity and send them homeward. Again the burn had to be carried out over the Moon's hidden side, out of contact with Earth. Again the burn was good. As contact was re-established, a message crackled through to mission control from the Moon: 'Please be informed, there is a Santa Claus.'

The danger was not over, however. Perhaps the most hazardous part of the mission was yet to come — re-entry into the Earth's atmosphere. On the early morning of 27 December, Apollo 8 was hurtling towards the atmosphere at a speed of nearly 25,000 mph (40,000 km/h). The service module was then jettisoned and the command module containing the crew angled heat-shield forward towards Earth. As it roller-coastered through the atmosphere its speed dropped rapidly. The heat shield glowed red-hot and began

Far left: Edwin Aldrin about to drop down onto the lunar surface to become the second man on the Moon on 20 July 1969. Armstrong had taken his 'one small step' minutes earlier.

Left: On the flat dusty expanse of the Sea of Tranquillity, Edwin Aldrin poses for the most famous of all space photographs. Reflected in his gold-tinted visor are photographer Neil Armstrong and the lunar module.

Below: With the planting of a human footprint on another world, a new era in Man's history begins. No longer shackled to our home planet, we have taken the first steps that may take us ultimately to the stars.

to melt, as it should, to dissipate the frictional heat and protect the astronauts inside.

At an altitude of about 33,000 feet (10,000 meters) the small drogue parachutes opened to slow Apollo's descent, followed by the three main parachutes at 10,000 feet (3000 meters). The splashdown came in darkness within 3 miles (4.5 km) of the recovery ship. At first light the astronauts were whisked aboard with much jubilation.

Acting NASA Administrator Thomas Paine described the mission as 'one of the great pioneering efforts of mankind. It is not the end but the beginning'. It proved indeed to be the beginning of the final assault on the Moon that would climax the following year in the first lunar landing.

The LM proves itself

Early in the New Year, 1969, the crew members were announced for Apollo 11, which would attempt the first Moon landing. They were to be civilian Neil Armstrong, ex-test pilot for the X-15 rocket plane; Air Force Colonel Edwin 'Buzz' Aldrin; and Air Force Lieutenant Michael Collins. The launch was scheduled for July. But there were, to say the least, still one or two things that needed sorting out.

Following the success of Apollo 8, NASA knew that the Saturn V, the Apollo CSM and the flight plan were faultless. But the lunar module had still to be flight tested, and the rendezvous and

Right: Dramatic lighting accentuates this Apollo 11 scene as Aldrin sets up a panel to record the composition of the solar wind – the stream of electrified particles that course through space from the Sun.

Below: Back safely inside Eagle, with 48 pounds (22 kg) of Moon rock and soil after his two and a half hour EVA, Neil Armstrong can afford to smile.

50

Inset left: Eight days three hours after leaving Kennedy on the trip of the century, the Apollo 11 astronauts splash down in the Pacific, south-west of Hawaii. Here, two of the crew sit in the life-raft, while the other helps one of the recovery team to close the spacecraft's hatch. Note the masks and the clothing the astronauts are wearing, known as biological isolation garments. This is worn as a precaution to prevent the transmission of any possible Moon germs from the astronauts.

Left: The Apollo 11 command module is hauled aboard the recovery ship USS Hornet. Its base heat shield, which bore the brunt of the heating during the 25,000 mph (40,000 km/h) re-entry, is scorched and scarred.

Right: On board USS Hornet the astronauts are transferred to a mobile quarantine facility to keep them isolated. It is a time for rejoicing, and President Richard Nixon, on hand to welcome them home, joins in enthusiastically. First man on the Moon Armstrong is on the left inside the quarantine facility; Edwin Aldrin is on the right; with Michael Collins in between.

docking procedures that would be necessary en route for, and in orbit around the Moon had to be rehearsed. These were to be the objectives of the next two missions, Apollo 9 and Apollo 10.

Apollo 9 soared off the pad on 3 March for a 10-day trip that was to be near-perfect from lift-off to splashdown. It was the first time the whole Apollo spacecraft, CSM and LM, had flown. The CSM with command (crew) module uppermost had to be on top of the launch vehicle so that the crew could escape in an emergency during launch. The LM was housed in the third stage of the Saturn V beneath it. For the trip to the Moon, however, the CSM would have to dock with the LM. The first main task of the Apollo 9 crew in the CSM, therefore, was to separate from the third stage rocket that accompanied them into orbit. The CSM was then turned round and docked with the LM to pull it clear from the third stage. So far, so good. Next the crew fired the service module's engine to

see what effect it would have on the CSM-LM combination. Everything went fine.

On the fifth day of the mission the time came to rehearse the rendezvous and docking procedures that would be necessary in lunar orbit. So two of the astronauts crawled into the LM, undocked from the CSM, and fired the LM's engines to separate the craft by several miles. Later, they pulled further away until they were some 125 miles (200 km) behind the CSM and out of visual contact. Using its own radar and instruments, the LM chased and successfully docked with the CSM with pinpoint accuracy. The LM had proved itself a spaceworthy vehicle.

Apollo 10's task, when it came to its launch, was to repeat what Apollo 9 had done only en route for the Moon and in lunar orbit. In short, the astronauts would take part in a dress rehearsal of the Moon-landing mission, without actually landing. Apollo 10's 8-day

Flight of the Eagle

After a flawless mission, Apollo 10 returned to Earth on 26 May. By then the Apollo 11/Saturn V vehicle had already been installed on the launch pad, and the world had begun counting down to a lunar landing.

The official countdown started on 10 July. While technicians feverishly worked on the 36-story-high rocket on the launch pad, the astronauts were having their final briefings. By the evening of 15 July, the roads around the Space Center, the river fronts, the beaches were packed with an estimated one million spectators, hoping to witness the beginning of the greatest adventure of all time.

Astronauts Armstrong, Aldrin and Collins entered the Apollo 11 CSM, code-named *Columbia*, at about 7 am EDT (Eastern Daylight Time). Lift-off was scheduled for 9.32 am. As the last minutes of the countdown ticked away, the tension inside and outside the launch center became electrifying. Eight seconds before zero, the Saturn V's five rockets roared into life and built up thrust. At 9.32 precisely the hold-down clamps were released and the rocket began to climb from the pad. The throbbing thunder of the rocket exhausts drowned the yells, cheers and prayers of the spectators as they willed the three astronauts to the surface of the Moon.

From then on, Apollo 11 followed the procedures practiced on earlier missions. Just after blasting into a lunar trajectory, the crew separated *Columbia* from the third stage and docked with the LM, code-named *Eagle*. In that configuration they journeyed uneventfully to the Moon. On 19 July they entered lunar orbit. The following day Armstrong and Aldrin transferred to *Eagle* and separated from *Columbia*. Firing *Eagle*'s descent engine, they swooped down towards the desolate lunar plain known as the Sea

Left: On the second Moon landing mission, Apollo 12, in November 1969, the fragile lunar module Intrepid *has just separated from the CSM* Yankee Clipper *and, with landing legs extended, is heading for a landfall on the Ocean of Storms. How beautiful the stark lunar landscape looks.*

Above: The Apollo 12 landing site shows the typical flat landscape of a mare region, created millions of years ago by a vast lava flow, perhaps triggered by the impact of a giant meteorite. As elsewhere on the Moon, the surface is covered with dust. The astronaut here is deploying components of ALSEP, the Apollo lunar surface experiments package, during the first EVA period.

Right: On their return journey back to Earth, the Apollo 12 astronauts photograph this artificial solar eclipse as the Earth moves directly in front of the Sun.

mission, which began on 18 May, proved highly successful. The LM separated from the CSM in lunar orbit and swooped tantalizingly to within 50,000 feet (15,000 meters) of the Moon's surface. Then it had to climb back to rendezvous with the mother ship. How frustrating it must have been for the two astronauts inside, Thomas Stafford and Eugene Cernan, to come so near to making the first lunar landing. That honor, however, was reserved for others.

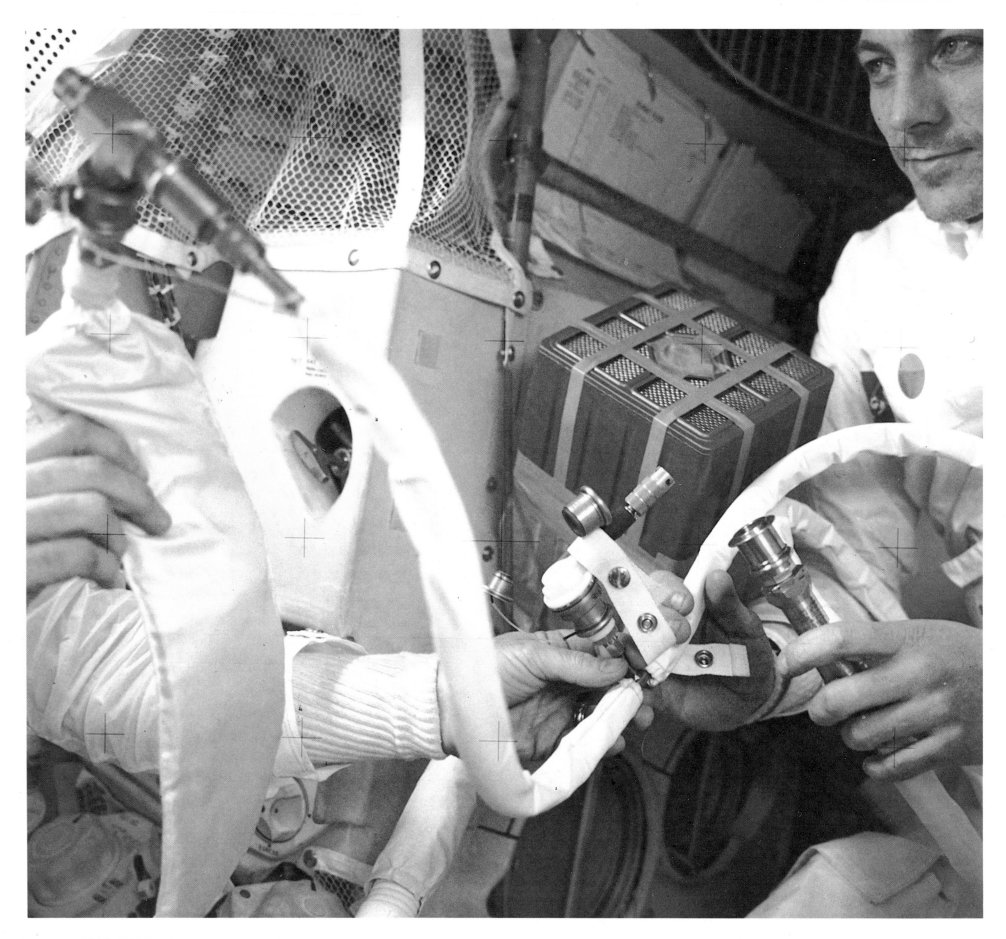

One giant leap

Some six hours later the two astronauts had suited up for the next great challenge – walking on the Moon. They depressurized *Eagle*, and first Armstrong emerged from the hatch and made his way down the descent ladder. He switched on a camera that recorded his actions for posterity. At 10.56 pm EDT he stepped down on to the lunar surface with his left foot. 'That's one small step for a man, one giant leap for mankind,' said Armstrong, watched by the biggest ever TV audience of around 600 million back on his home planet.

The fantasies and dreams of Earth-bound humanity throughout the ages; the visions of such writers and scientists as the Greek Lucan 1700 years ago; the English Bishop Goodwin and France's Cyrano de Bergerac, in the 17th century; Jules Verne and H.G. Wells in the 19th century; Tsiolkovsky, Goddard, Oberth and Von Braun – had become reality. And American people had achieved the goal set for them by President Kennedy. They had landed a man on the Moon before the decade was out.

Aldrin soon joined Armstrong on the lunar surface, and together they set up an experiment to detect the solar wind. They then planted a US flag in the soil, stiffened with wire to make it 'fly' in the airless world. President Nixon put through from the White House 'the most historic telephone call ever made' to congratulate them. Afterwards the two pioneering moonwalkers collected rock and soil samples and set up a laser reflector and a seismometer to detect 'moonquakes'. By 1.10 on the morning of the 21st both astronauts were safely back inside the lunar module, exhausted but elated; their moonwalk had taken just over two and a half hours.

Left: Human ingenuity triumphs over adversity during Apollo 13's ill-fated attempt at a Moon landing in April 1970. The crew use the lunar module Aquarius *as a liferaft when an explosion knocks the CSM* Odyssey *out of action. John Swigert and another astronaut display the makeshift air purifier that helps prevent the build-up of suffocating carbon dioxide.*

Above: President Richard Nixon flew to Honolulu to greet the Apollo 13 astronauts after their safe splashdown in the Pacific and to present them with the Medal of Freedom, the nation's highest civilian award. Here he talks with command module pilot John Swigert.

Right: The Apollo 14 lunar module Antares *reflects the strong sunlight at its landing site in the hilly Fra Mauro region of the Ocean of Storms in February 1971. The astronauts perform two EVAs totaling 7 hours 23 minutes before rejoining their colleague in the CSM* Kitty Hawk.

of Tranquillity. Seeing that *Eagle* was heading towards a crater the size of a football field, Armstrong took over manual control of the LM and aimed for a smoother landing site.

Seconds before 4.18 EDT on 20 July 1969 *Eagle* touched down. 'Houston,' reported Armstrong, 'Tranquillity Base here. The *Eagle* has landed'.

'Roger Tranquillity,' replied Houston, 'we copy you on the ground. You got a bunch of boys about to turn blue. We're breathing again. Thanks a lot.'

Far left: At the Kennedy Space Center technicians stow the collapsible lunar roving vehicle, or Moon buggy, aboard the Apollo 15 lunar module. This ingenious $15 million lunar jeep will allow the astronauts to roam more widely and make the most of their limited time on the Moon.

Left: In August 1971 the astronauts in the Apollo 15 lunar module Falcon *take this picture of the CSM* Endeavour *just before docking with it after their lunar exploration. The picture shows the open scientific instrument bay (SIM), which holds instruments for scanning the lunar surface from orbit.*

Below: Apollo 16 begins the penultimate Moon-landing mission with a characteristically spectacular lift-off on 16 April 1972. Aboard is John Young, making a record fourth space flight.

time they were kept in quarantine, together with the samples they had brought back from the Moon, at the Lunar Receiving Laboratory at Houston. Then they and the samples were declared germ-free. The Moon seemed to be sterile. They emerged from quarantine on 10 August, and two days later the celebrations began – tickertape parades, motorcades, galas, dinners, speeches – and in September an exhausting world tour. Such was the price of fame.

The unlucky 13th

Ten more astronauts followed in Armstrong's and Aldrin's footsteps over the next three and a half years, on Apollo missions 12, 14, 15, 16 and 17.

The excitement of the first Moon landing had scarcely evaporated when Apollo 12 set out to make the second lunar landfall, on 14 November 1969. It was a spectacular lift-off, accompanied by bolts of lightning, as the Saturn V and its flaming tail apparently acted like a lightning conductor amid the storm clouds over the Cape. The mission itself was flawless. The lunar module made a pinpoint landing on the Ocean of Storms within walking distance of a Surveyor probe which had soft-landed there in 1967. Moonwalkers Alan Bean and Charles Conrad walked over to it and removed pieces to take back to Earth for inspection.

Apollo 13 took off on 11 April 1970. The initial part of the journey proceeded with accustomed smoothness. But two days later, 55 hours 55 minutes after lift-off, the mission became the unlucky 13th. The spacecraft was over 200,000 miles (320,000 km) from home, and accelerating towards the Moon. The crew had just completed a telecast to Earth and were clearing up. John Swigert was in the command module (code-named *Odyssey*), Fred Haise

Some thirteen hours later they returned to lunar orbit in the upper stage of *Eagle* to rendezvous with Collins in the CSM *Columbia*. *Eagle* was then jettisoned. On the thirtieth orbit of the Moon, *Columbia*'s engine fired to boost the astronauts back to Earth. The return flight was, like the outgoing one, uneventful. Splashdown occurred just after midday on 24 July in the Pacific Ocean south-west of Hawaii.

When the crew emerged from *Columbia*, they were given biological isolation suits to wear, just in case they had brought back any lunar germs. They were then put in isolation in a mobile trailer on the recovery ship, where President Nixon was waiting to greet them. With hyperbole that could perhaps be forgiven, he said. 'This is the greatest week in the history of the world since the creation.'

The American people and the world press had to wait for another two weeks to meet the lunarnauts in person. Until that

Left: During their exploration of the Descartes Highlands region of the Moon, the Apollo 16 astronauts visited Plum Crater (left). Here Charles Duke is boring into the soil to obtain a core sample. Note the lunar rover on the opposite side of the crater, in which the astronauts set a new lunar speed record of 11.2 mph (18 km/h), downhill.

Right: As Apollo 16 returns home after successfully completing a fifth lunar landing, the astronauts photograph the full Moon. Note that it presents a slightly different aspect than usual.

was in the lunar module (*Aquarius*), with James Lovell in between. Suddenly there was a loud bang, the spacecraft vibrated, and within seconds the master alarm sounded.

Mission control back on Earth were stunned when Swigert radioed through: 'Okay Houston, we've had a problem here.' It was a masterly understatement. What had happened was a potential disaster. A liquid oxygen tank in the service module had exploded, destroying the fuel cells that supplied power to the spacecraft and cutting off the oxygen supply. The service module, including its propulsion motor, was dead. The command module had a back-up battery pack, but that would be needed for re-entry. In any case it had a life of only 10 hours, and Apollo 13 was 87 hours from home.

The crew's salvation rested with the lunar module *Aquarius*. For the next three days they had to rely on its limited power supply, its oxygen, and its engines to get them back home. They fired the LM's

descent engine to change Apollo's trajectory into one that would swing them around the Moon and direct them back to Earth. The return journey was uncomfortable as the temperature in the crippled craft progressively fell. A build-up of carbon dioxide was also a problem until the crew rigged up a system of hoses that circulated the cabin air through the command module's carbon dioxide scrubbers.

On 17 April Apollo 13 was within sight of home. The crew transferred to the command module and cut loose the lifeless service module. As the structure drifted away they could see the extent of the damage. A whole side had been blasted away. Next they jettisoned their lunar module 'liferaft' and prepared to hit the Earth's atmosphere traveling at thirty-five times the speed of sound. Within a quarter of an hour, they were rocking gently in a Pacific swell. The agony was over. Said President Nixon afterwards: 'You did not reach the Moon, but you reached the hearts of millions of people on Earth.'

Lunar roving

The next launch, of Apollo 14, was scheduled for October, but investigation of Apollo 13's near-disaster and the ensuing modifications to the service module, caused postponement until 31

Left: On the final Apollo mission in December 1972, geologist Harrison Schmitt uses a kind of 'rake' to collect rock chips of certain sizes. During a record 22-hours of EVAs, Schmitt and fellow lunarnaut Eugene Cernan collect no less than 250 pounds (113 kg) of soil and rocks.

Above: With the Taurus Mountains as a backdrop, Eugene Cernan takes the invaluable lunar buggy for a spin. Later it will transport Cernan and Schmitt and their equipment for a 22 mile (35 km) geological safari.

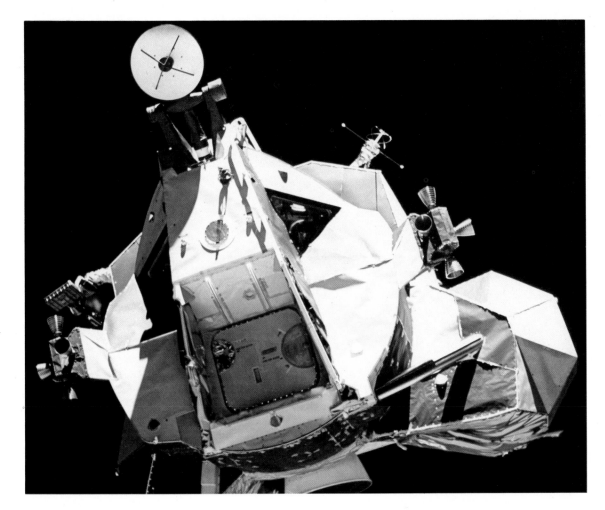

range on the edge of the Sea of Rains. The 15,000 feet (5000 meter) peaks provided a stunning backdrop to the landing site. The astronauts journeyed to nearby Hadley Rille, a 1200 foot (300 meter) fracture in the lunar surface.

Apollo 16 made landfall on the Cayley Plains, a highland region near the crater Descartes, in April 1972. It was chosen because the highlands are thought to be part of the Moon's original crust. The mare regions, by contrast, were formed by more recent lava flows.

The final mission to the Moon, Apollo 17, started spectacularly, just after midnight on 7 December 1972. This first night launch of the great Saturn rocket was observed as far away as the state of Georgia. The Apollo 17 landing site was located in a rugged mountain valley called Taurus-Littrow, on the south-east boundary of the Sea of Serenity. It was thought to be particularly interesting geologically, and a geologist, Harrison Schmitt, was included in the flight crew. Schmitt made the most of his unique scientific opportunities as he examined the landscape, sampled the soil, and hacked with a geological hammer at the massive rocks.

Before Schmitt and his fellow moonwalker Eugene Cernan returned to their lunar module, there was one last act to perform. Cernan read aloud the inscription on a plaque attached to the descent stage of the lunar module, which would remain on the Moon. The plaque showed a picture of the two hemispheres of the Earth, with the Moon in between. 'Here man completed his first exploration of the Moon, December 1972. May the spirit of peace in which we came be reflected in the lives of all mankind.'

The men from planet Earth left the Moon at 5.55 pm EST on 14 December 1972, an event witnessed by the TV camera on their lunar roving vehicle. Within three days, they were home. So ended the most remarkable episode in human history. No-one knows when human feet will once again be planted in the lunar soil.

The scientific harvest

The twelve astronauts who left their footprints in the lunar dust explored the Moon for a total of nearly 170 hours and roamed the surface for nearly 60 miles (100 km). Statistically the treasure trove of data they brought back from the Moon is overwhelming: 850 lb (385 kg) of soil and rocks; 50 core samples; 30,000 photographs, taken from the ground and from lunar orbit; 20,000 reels of tapes containing geophysical data. Even today samples and data wait to be analysed thoroughly.

Left: Astronaut Harrison Schmitt, the Stars and Stripes and the distant Earth are squeezed into this classic Apollo photograph.

Above: Seen from the Apollo 17 CSM America *is the lunar module* Challenger *coming in to dock after its ascent from the Moon. The strictly functional shape of the lunar module is clearly evident.*

Right: Seconds before splashdown, the Apollo 17 command module is lowered gently down to the sea by its three main parachutes. It is the end of the most exciting four years in the history of human exploration. The Apollo Moon-landing program is over, and life on Earth will never be quite the same again.

January 1971. Apollo 14 was also retargeted to Apollo 13's original destination, a hilly site near the crater Fra Mauro on the edge of the Ocean of Storms. The major tasks for astronauts Alan Shepard and Edgar Mitchell during the first of two EVAs was to set up the self-contained scientific station known as ALSEP (Apollo lunar science experiments package).

On the second EVA they went moonwalking, exploring the surrounding landscape with their equipment carried on a two-wheeled 'golf cart', or to give it its official name, the modularized equipment transporter. Perhaps 'golf cart' was apt because Shepard, just before he re-entered the lunar module, produced three golf balls. Using the handle of one of the geological tools as a driver, he swung at each of them. He missed the first but clouted the other two. In the one-sixth lunar gravity, they went, Shepard said, 'miles and miles and miles'.

The three following missions, Apollo 15, 16 and 17, were longer, more ambitious and highly successful. On each, the moonwalkers performed three EVAs, during which they set up ALSEP stations. They were also able to explore much farther afield than their predecessors thanks to their unique transport, the $15 million lunar roving vehicle, or Moon buggy. Powered by electric motors on all four wheels, the buggy had a maximum speed of 10 mph (16 km/h).

The scenery on the last three missions was spectacular. In July 1971 Apollo 15 landed in the foothills of the Apennine mountain

Scientifically the harvest reaped from the Apollo missions has been abundant. We at last know exactly what our nearest celestial neighbor is like. This dead world of rugged highlands and lava-filled 'seas', or maria, is covered with a kind of top soil (regolith) that has the consistency of newly plowed Earth soil.

The rocks are similar in some respects to the basalts and breccias found on Earth. But their composition differs from that of terrestrial rocks. They contain no new elements, but are made up of much larger amounts of elements relatively rare on Earth, such as chromium, titanium and zirconium. Studies from orbit showed that radioactive elements such as thorium and uranium are also more abundant. They also revealed curious concentrations of super-dense matter beneath some of the seas, named mascons (mass concentrations).

The ALSEP scientific stations left behind on the Moon transmitted their results back to Earth, reporting about the Moon's magnetic field, about charged particles and radiation, and about 'moonquakes'. Some are still reporting back today, though no-one is listening any more. The seismometers in ALSEP reported that the Moon is geologically quiet, as expected. There are only a few thousand ground tremors a year, as opposed to a million or more on Earth. The Apollo astronauts created several artificial 'quakes' when they deliberately crashed their jettisoned lunar modules on the surface. When this happened, the shock waves created persisted for hours, making the Moon seismically 'ring like a bell'.

The differences in composition and structure between the Earth and Moon suggest that the Moon was not once part of the Earth as some scientists believed. The age of the lunar rocks tends to confirm this. They are very old indeed. Most are over 3 billion years old, and some are over 4 billion. This is getting close to the age of the Earth and the solar system (4.5 billion years). So it seems certain that the Moon was formed at the same time as the Earth, and was perhaps captured by the Earth's gravity in the far distant past.

Because it's there

The total cost of the Apollo project is reckoned to have been in the region of $25 billion. Posterity will judge, ultimately, whether it was worth such cost. Many would say that the money could have been better spent. Be that as it may. But one thing is certain, mankind would have gone to the Moon one day anyway. It is a cosmic Everest to be conquered simply 'because it's there'.

A bonus from the Apollo flights was that they gave us a chance to see our planet in perspective – a vibrant, warm, colorful oasis of life in the hostile abysmal desert of space. Poet Archibald MacLeish summed up the feelings of many when he wrote after Apollo 8's pathfinding voyage:

'To see the Earth as it truly is, small and blue and beautiful in that eternal silence where it floats, is to see ourselves as riders on the Earth together, brothers on that bright loveliness in the eternal cold – brothers who know they are truly brothers.'

Right: This sample of rock brought back from the Moon is about 4 inches (10 cm) across. It is volcanic in origin and is almost pumice-like in structure. Of the 850 pounds (385 kg) of soil and rocks returned by the Apollo astronauts, most still await analysis even today.

Far right: The various constituent minerals in a thin slice of Moon rock show up under the geological microscope. The minerals found on the Moon are significantly different from those found on Earth, reflecting different conditions of formation.

Chapter 4

New Directions

The year 1972 marked the end of the Apollo Moon-landings but not the end of the Apollo program. Even as Apollo 17 was on its way back from the Moon in December, preparations were well advanced for another spectacular mission using Apollo hardware. It was to be a laboratory in space, a place where astronauts could work for long periods in orbit; in effect an experimental space station. Named Skylab, the space laboratory was scheduled for launch in May 1973.

There was also another Apollo spin-off in the planning stages. In May 1972 at the US-USSR summit a five-year agreement had been signed concerning 'cooperation in the exploration and use of outer space for peaceful purposes.' The agreement gave the go-ahead for a joint US-USSR space flight, which would culminate in a link-up in orbit between Apollo and Soyuz spacecraft. It became known as the Apollo-Soyuz Test Project (ASTP). The great space rivals would join forces for an international flight in the mid-1970s. It was a good direction to be heading.

NASA was looking much farther ahead, however. Technologically, the Apollo hardware was becoming outdated. The huge Saturn rockets were an expensive way of launching objects into space. They were expendable – they could only be used once. Take the Saturn V for example. Three hundred and sixty-five feet (111 meters) of expensive hardware went up, and only the blackened cone of the command module, a few feet across, came back down. And even that couldn't be used again.

What was needed was a manned reusable launch vehicle that could be used over and over again, a space truck that could shuttle back and forth between Earth and space. This was the proper and most efficient way of transporting payloads into orbit. Out of this need arose the space shuttle. The year 1972 was a critical year for that too, for in January President Nixon gave formal approval to its development.

The origins of Skylab

As mentioned earlier, the Apollo program originally embraced the development of hardware not just for Moon-landings, but also for Earth-orbital activities. Priority was given to the Moon-landing aspect after Kennedy's May 1961 exhortation to Congress. The Earth-orbital side of the program was not entirely neglected, but the increasing financial demands of Mercury, Gemini and Apollo development left little money over for post-Apollo projects.

As far back as 1965, NASA outlined to President Johnson what it

These pages: On 14 May 1973 the last Saturn V to be launched lifts off launch Complex 39 bound, not for the Moon this time, but for Earth orbit. It carries the Skylab space station. The lift-off is perfect, or so it seems.

The Skylab program emblem.

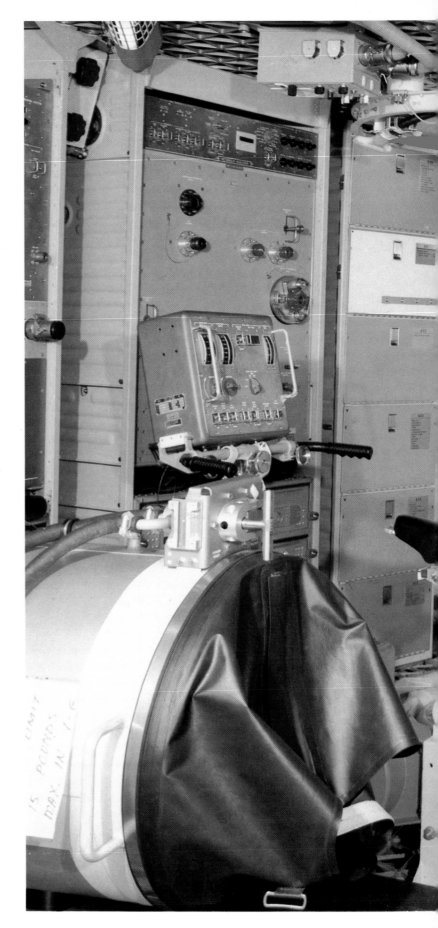

Above: The Skylab program emblem. It depicts the space station in its intended configuration in orbit, with two solar panels projecting from the sides of the large cylinder, which is the orbital workshop.

Right: This life-size and realistic mock-up of Skylab's orbital workshop is now on display at the Johnson Space Center at Houston. It shows part of the crew quarters. The mannequin is seated at the food table. In the left foreground is a lower body pressure device used in Skylab experiments, and behind that a bicycle ergometer for medical studies and crew exercise.

considered to be the priorities for the post-Apollo era. They included exploration of Mars by robot vehicles; long-stay missions to the Moon to conduct more extensive studies and last in order of priority a manned Earth-orbital program. Budget cuts, however, eventually forced what had come to be called the Apollo Applications Program (AAP) to concentrate on the latter.

The main object of the AAP was to support long-duration orbital missions, during which the astronauts would carry out wide-ranging experiments in space science and technology. Going for cheapness, NASA recommended the use of spent rocket stages to support a variety of experimental modules. This idea originated from von Braun's design team at the Marshall Space Flight Center. The rocket stage in question was the S-IVB, the upper stage of both the Saturn IB and the Saturn V launch vehicles already under development.

Skylab plans
On 26 January 1967 George Mueller, NASA's Associate Administrator of Manned Spaceflight, announced at a press conference that AAP missions would start the following year. Initially, a Saturn IB would be launched into Earth orbit, with its

S-IVB upper stage. Other launchings would put up a crew of three astronauts in a modified Apollo CSM and a Multiple Docking Adapter (MDA). In orbit the S-IVB would be dried out of fuel and linked up with the MDA and the Apollo CSM. It would then be fitted out as a habitable workshop with equipment carried up in the other units. This spent-stage concept was quite reasonably known as the 'wet' workshop.

There would be several visits to the workshop by different three-man crews, and flights up to 56 days were envisaged. Later there would be other modules launched, including a solar observatory, called the Apollo Telescope Mount. This would be attached to the workshop in orbit. Missions up to a year would follow, as would other modular workshops.

The day after the press conference, however, all plans for post-Apollo activities were shelved. This was the day that Grissom, White and Chaffee burned to death on the launch pad of Complex 34 at Cape Canaveral. Immediately, many of the resources allocated to the Apollo Applications Program were switched to redesigning the Apollo command module and catching up with Moon-landing schedules.

The severe cut-back in finances led to a re-think of the AAP. It would not, as intended, run concurrently with the Moon-landing program, but follow it. It would have to be restricted to a single workshop, rather than several. It was also becoming clear that the idea of a 'wet' workshop using a spent rocket stage had many drawbacks. How much better it would be if the workshop could be launched 'dry' and fully equipped. However, this would require the services of a Saturn V rocket to launch it.

NASA Administrator James Webb would not initially allow one to be diverted from the Apollo Moon-landing program. However, when Apollo 11 returned safely from the Moon in July 1969, approval for the use of a Saturn V was given. And a month later the decision was taken to switch to a 'dry' workshop. In February 1970, when the AAP was renamed Skylab, the mission planning was well advanced and the final form of the orbiting structure had been settled. Launch was scheduled for 1973.

Workshop in orbit

The main part of Skylab was a modified S-IVB rocket stage of a Saturn IB. This was called the orbital workshop (OWS). It provided the crew quarters, experiment compartment and storage for most of the expendables such as food and water. The outside was protected from impact by space particles by a meteoroid shield, which also provided thermal protection. Attached to the sides of the OWS was a pair of solar cell arrays as big as barn doors, designed to supply the workshop with some 10 kilowatts of power.

The OWS was essentially a two-story structure. The lower level comprised the crew quarters, which were split into four sections. The astronauts' living area was the wardroom, which contained a galley for food preparation and a table for eating. Adjacent to the wardroom were Skylab's toilet facilities, technically known as the waste management compartment. Next door was the sleeping compartment, where the astronauts had separate sleeping quarters provided with sleeping bags stretched between ceiling and floor. Finally there was an experiment compartment with a variety of apparatus such as a rotating chair and a bicycle ergometer.

The upper level of the OWS, called the forward compartment,

was a major storage area for food, clothing and experimental equipment. The astronauts' spacesuits were also stored there. Above this compartment was the end dome of the rocket casing.

The whole of the living space was contained within what was originally intended as the liquid hydrogen tank for the rocket stage. Beneath the crew compartment level was the tank intended for liquid oxygen. In Skylab this was used as a waste tank to store trash.

Attached to the forward dome of the OWS was an airlock module (AM). Astronauts could isolate this from the rest of the craft and depressurize it when they needed to perform EVAs. The AM also doubled as a control center for Skylab, particularly for its life-support and communications systems. It held the nitrogen and oxygen tanks that supplied the atmosphere in Skylab, which was maintained at a low 5 pounds per square inch (0.35 kg/cm^2), about a third of the atmospheric pressure on Earth.

Left: Saturn V, with Skylab installed in its upper stage, awaits its launch, bathed in the soft ruddy glow of a Florida sunset.

Right: When the first team of Skylab astronauts arrive at the space station, they have to make good the damage caused by the launch. They erect a sunshade to stop the station overheating and (above) free a jammed solar panel. The astronauts in the picture are Charles Conrad (top) and Joseph Kerwin.

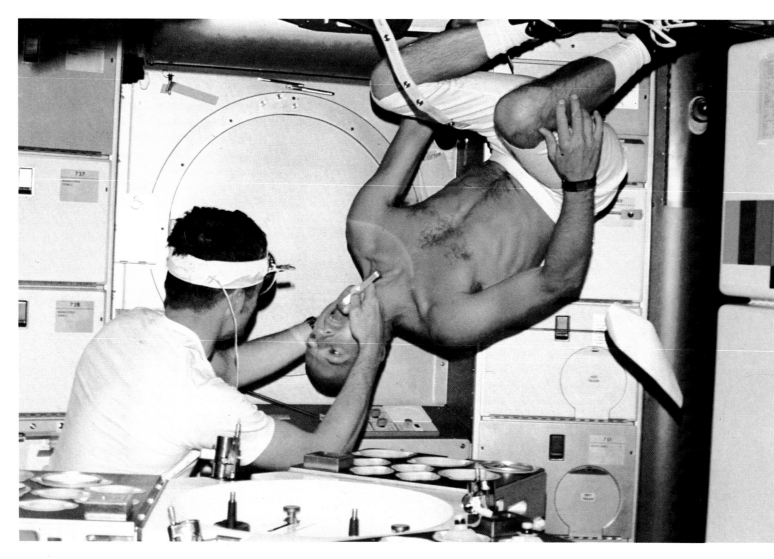

The AM was also linked to the multiple docking adapter (MDA). At the end of this was the docking port through which the astronaut crews entered the orbital complex from their Apollo command and service module (CSM), in which they were ferried up from Earth. A second emergency docking port was provided in the side of the MDA.

The MDA also housed the main control console for the Apollo Telescope Mount (ATM) solar observatory, which was mounted on the MDA/AM modules. The ATM was a windmill-like structure with four solar panels extending out from a central hub, in which a package of eight solar telescopes and sensors was located. These solar 'panels' were designed to generate nearly half of Skylab's electrical power.

The complete Skylab cluster, including the CSM docked at one end, measured some 119 feet (36 meters) long, had a maximum diameter of 27 feet (8 meters) and weighed 100 tons (90 tonnes). Its total volume of some 12,000 cubic feet (330 cubic meters) was similar to that of a small house.

A Russian Salyut

As planned, the Skylab mission would begin with the space station, with all its solar panels neatly folded by its sides, being launched by a Saturn V rocket. It would take the place of Saturn V's normal third stage. In orbit the solar panels would deploy, and Skylab would be ready to receive the first visit by a crew of three astronauts, who would be launched by a Saturn IB the day after. The first crew would man the space station for 28 days then return. Later two further crews would visit Skylab for two 56-day periods. But this was not quite what happened.

Launch day for the unmanned Skylab was set for 14 May 1973. The countdown proceeded smoothly, and precisely on time the Saturn V thundered into the heavens. The United States would soon have its space station.

It would not, however, be the first space station. As if to prove that the space race had not finished on the Moon, two years earlier the Russians had launched their first space station, Salyut 1. The Salyut had been very much smaller than Skylab, only some 45 feet (14 meters) long. And its time in orbit had been attended first by triumph then by tragedy.

On the second visit to Salyut, in the Soyuz 11 spacecraft, cosmonauts Georgi Dobrovolsky, Viktor Patseyev and Vladislav Volkov had spent a record 24 days in space. They had returned to Earth on 29 June 1971, landing, as cosmonauts usually do, on land in a spherical re-entry capsule. But as the recovery team rushed

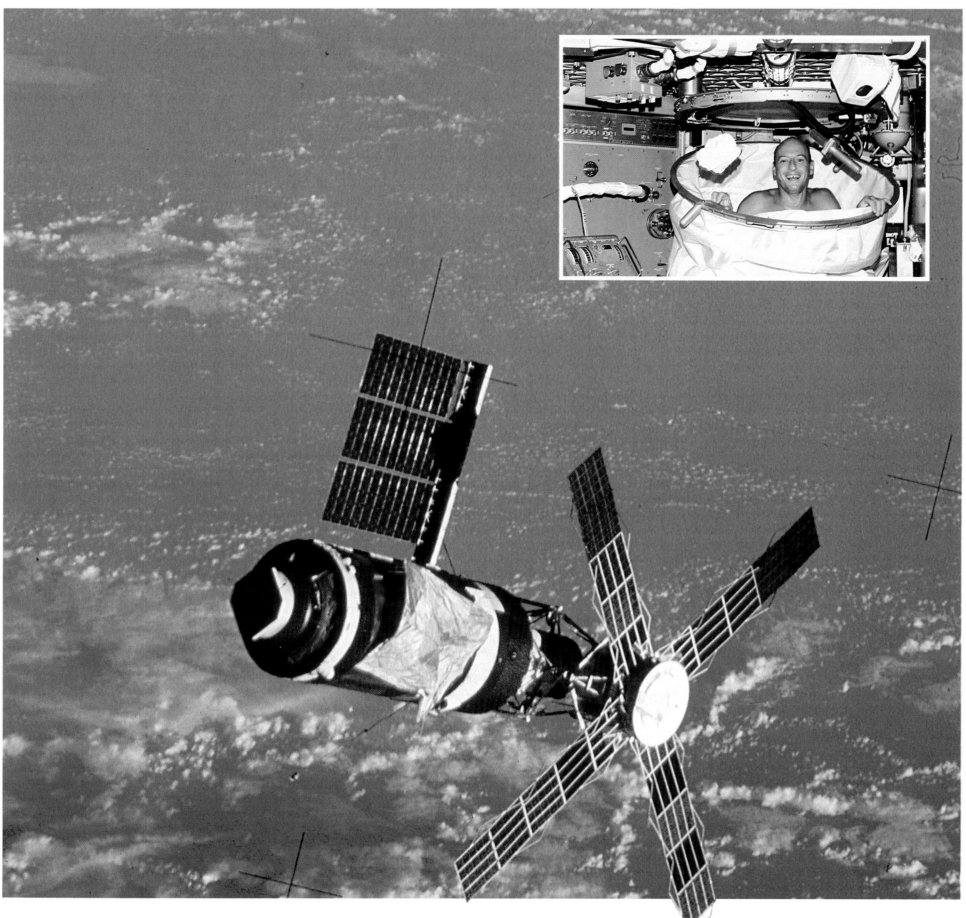

towards it, no-one had come out. Inside, the cosmonauts were dead. Speculations that such a long stay in space had weakened them to the point of death proved unfounded. They had been killed by spacecraft depressurization after an equipment malfunction.

Plans go wrong

NASA knew that the Skylab soaring up to orbit on 14 May 1973 was a much more sophisticated space station than Salyut and once in orbit would revolutionize space science. However, once Skylab had reached its intended 270 mile (435 km) orbit, things began going wrong. According to the signals received from Skylab, the solar panels on the sides of the OWS had not deployed and were not generating power. The ATM solar panels were working, but not providing enough power for Skylab's needs. There was also no indication that the meteoroid shield had deployed. Then the temperature in Skylab started to rise.

On examining the data from the beginning of the mission, it was

Above: The second Skylab team take up occupation of the space station in July 1973. Here Owen Garriott tucks into a square meal, using a knife, fork and spoon. Skylab astronauts are the first to make use of conventional cutlery for eating.

Right: Owen Garriott performing a spacewalk. He is seen near the hub of the X-shaped Apollo Telescope Mount, having just deployed an experiment (the blue package near his head) to collect interplanetary dust particles.

found that meteoroid shield deployment had occurred at 63 seconds into the flight, while the rocket was accelerating at its peak. This was thought at the time to be an erroneous signal. But it wasn't. The shield *had* deployed and had been ripped off. For good measure, it had also carried away one of the OWS solar panels and jammed the other. With no shield to protect it from the Sun, Skylab had begun to overheat.

The manned launch planned for the next day was postponed, while NASA pondered what to do. The first thing they did was to try to reduce the temperature inside the workshop as much as possible. Over the next few days they changed the orientation of Skylab to the Sun, and managed to bring the temperature down from about 48°C (116°F) to 32°C (90°F). Had it gone up much higher, instruments could have been affected, and plastic insulation could have started melting and releasing poisonous gas. The workshop would have become permanently uninhabitable.

Running repairs

It became clear that there was nothing else anyone could do on the ground to improve Skylab's condition. The decision was therefore made to go ahead with the Skylab 2 manned mission, and hope that the astronauts could carry out repairs on the spot.

The first priority was obviously going to be to rig up a sunshade where the meteoroid shield had been ripped off. Troubleshooters at the Johnson Space Center began working out ways of erecting one, while Skylab 2 astronauts Charles Conrad and Joseph Kerwin began practicing erection procedures in the giant water tank at the Marshall Space Flight Center, used for simulating weightless conditions.

The sunshade evolved into a kind of parasol of reflective plastic sheet on nylon ribs. It was to be rolled like an umbrella, pushed through a airlock in the OWS and then opened out and pulled down into position.

Below: The second Skylab team (Skylab 3 mission) take some remarkable photographs of Earth with the Earth-resources experiment package (EREP) at non-visible wavelengths. This is an image of San Francisco and its environs. Note the great bank of fog moving in off the Pacific.

Right: This Skylab EREP picture shows the fertile farmland on both banks of the lower Mississippi River. Vegetation shows up red on this imagery.

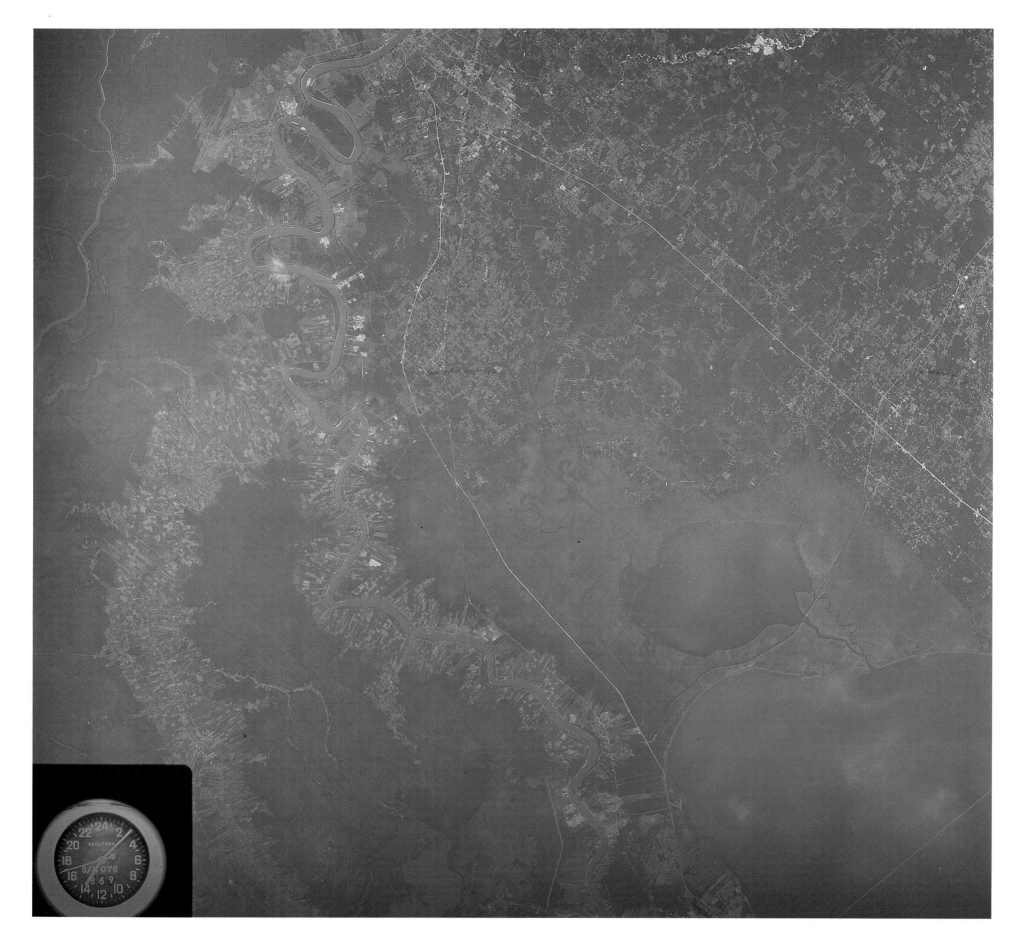

The Skylab 2 launch, on 25 May, took Kerwin and Conrad, and fellow crew member Paul Weitz, successfully into orbit. They confirmed the damage when they rendezvoused with Skylab. Weitz tried to unjam the remaining OWS solar panel by leaning out of the CSM with metal cutters fixed to a long pole. He had no luck, however. The crew, with some difficulty, then docked with the space station which by this time was still very hot but just habitable. Next day they successfully deployed the parasol through the airlock, and it worked perfectly. Within hours the temperature inside Skylab started to fall.

As the temperature fell, the crew began their scheduled work, switching on the solar instruments, filming the Earth, and starting the medical experiments with themselves as guinea pigs. On 7 June the crew addressed themselves to some unfinished business — the jammed solar panel. This time, working from the EVA hatch in the side of the airlock module, Conrad and Kerwin managed eventually to cut the restraining strap. The solar panel flipped open and began feeding desperately needed extra power into Skylab's electrical system.

The space station, in jeopardy since 63 seconds after lift-off, was now in a fit state to support not only the remainder of the first manned mission, but also the other two longer missions planned for later in the year.

Above: The Skylab 3 team also photographs this unusual view of the full Moon rising. Note the pronounced 'airglow' in the atmosphere.

Right: The Skylab 4 mission, the last visit to the station, begins on 16 November 1973. It is not to end for 84 days. It will be the longest any American will spend continuously in space until the advent of the space station in the 1990s.

Far right: This is the view that greets the Skylab 4 astronauts when they arrive in orbit. Like their predecessors, they have to carry out running repairs to keep the station fully operational.

The do-it-yourself talents of the astronauts, however, were called upon again and again to overcome more minor equipment malfunctions during the second half of their stay in orbit. In the best of 'fixit' traditions, they even resorted to none-too-gentle taps with a hammer to try to free a stuck relay in a battery regulator. While Conrad wielded the hammer, Weitz reported to Houston: 'There it goes. Boy is he hitting it. Holy cats!' Houston replied: 'It worked. Thank you very much gentlemen, you've done it again!'

Into the record books

On 18 June the Skylab 2 astronauts exceeded the 24-day orbital stay time of the Russians who had perished in Soyuz 11. They then went on to complete the planned 28 days.

Before they left the now comparatively healthy station, they switched off the necessary systems and 'mothballed' it for the period before the next crew arrived. Early in the morning of 22 June, they were back in the Apollo CSM and pulling away from their home of the past four weeks. They splashed down hours later in the Pacific, seemingly little the worse for wear from their experience. Subsequent medical checks confirmed this.

Consequently, flight plans were drawn up for a July launch for the Skylab 3 mission, which would last for 56 days, subject to periodic checks on the health of the astronauts and the condition of the space station. The crew of Skylab 3, Alan Bean, Owen Garriott and Jack Lousma, were also going to be called upon to do running repairs. Even while the Skylab 2 crew were in residence gremlins had been quietly at work in the space station. The parasol deployed over the damaged section of the OWS was deteriorating, and some of the gyroscopes that helped keep Skylab properly orientated were overheating and beginning to fail.

On 28 July the Skylab 3 crew, with a new sunshade and a 'six-pack' of gyroscopes, blasted off to make the second visit to the space laboratory. Within five hours they were docking with Skylab, but not before a problem arose with one of the 'quad' thruster units on their CSM, which began leaking propellant. The new crew moved quickly into their new home and began activating all the essential systems to get it back into fully working order.

Within hours the whole crew were complaining of 'space sickness': the dizziness and nausea and often vomiting that many astronauts experience when they first encounter weightlessness. The condition usually disappears after a few days, and it did with Skylab's new visitors. By 1 August they were beginning to get their 'space legs'.

Before they could get into their experimental stride, however, another problem arose. They were woken up on 2 August by a warning alarm, which indicated that another of the four quad thrusters of the CSM was leaking propellant. Like the previous one, it had to be shut down. Their CSM ferry had now lost half of its steering system. If it lost even one of the remaining two quads, it would be difficult to maneuver, and the astronauts' lives could be at risk when they eventually had to return to Earth.

Back on Earth the potential danger of the situation prompted NASA to prepare for a rescue mission, just in case, and work on the fourth mission's hardware was accelerated. Even so it would be early September before a rescue mission could be mounted, if it were needed. In the meantime there was nothing the crew in

Skylab could do except press on with the experiments and planned routines. They went on EVA to install a new sunshade over the parasol, and to replace six of the malfunctioning gyroscopes.

Early September came and went, as Skylab and the astronauts were cleared for a full 59-day mission. This came to an end when they splashed down on 25 September. Strangely enough, the astronauts were physically in far better shape than the first crew despite having spent twice as long aloft. The Skylab 4 crew of Gerald Carr, Edward Gibson and William Pogue lifted off for the final visit to Skylab on 16 November.

The space station was again ailing, this time with a leak in the cooling system. But the astronauts were carrying a repair kit with them, and the leak was plugged. They then settled down to the regular routine of tending experiments, eating, sleeping, exercising and occasionally complaining of the work load.

One month went by, Christmas came and went, as did the New Year. On 12 January the crew passed the Skylab 3's space endurance record of 59 days, which was officially their own stay time. But NASA decided to keep the mission going as long as possible so as to maximize the scientific return from the project, although by now the astronauts were spending more and more time simply keeping Skylab in a healthy state.

Not until 8 February, after 84 days in orbit, did Carr, Gibson and Pogue say farewell to the first great space laboratory and board their Apollo CSM, which ferried them back, without a hitch, to Earth. Incredibly, after traveling 34 million miles (55 million km) in space, they proved even fitter than either of the other crews!

The Skylab experiments

Skylab carried the largest collection of scientific hardware ever flown in space, the astronauts carrying out more than 250 scientific investigations in virtually every field that could benefit from the unique vantage point Skylab offered. Experiments were carried out in four major areas: the life sciences, solar physics, Earth observations and materials science.

Using themselves as guinea pigs, they monitored one another's bodily condition throughout their stay in orbit. This involved regularly taking blood and urine samples and participating in experiments on the cardiovascular system and the organs of balance, for example. Electrocardiographs monitored the crew's heart rates at intervals during rest periods and when they were pedalling on the bicycle ergometer.

The ergometer provided the main means of exercise for the crew. They also devised means of their own. A favorite one was to run around the walls of the forward dome of the OWS, using centrifugal force to help keep them going. Increased exercise was the major reason why the astronauts on the last two Skylab missions were so fit. Even after months spent in weightlessness, their bodies adapted to normal gravity within a few days.

A wealth of data was gleaned about the Sun from the telescope package in the center of the Apollo Telescope Mount. The astronauts, working in collaboration with hundreds of solar scientists on Earth, acquired in excess of 180,000 images of the Sun's disc over a wide range of wavelengths. They were fortunate to witness some of the biggest prominences — fountains of searing hot gas arcing above the Sun's surface — ever recorded. Skylab, in

Right: There is plenty of room in the upper chamber of Skylab's orbital workshop to perform interesting experiments, like how to balance a fully grown man on your finger tip! Here Edward Gibson is performing the trick with an impassive Gerald Carr. In zero-gravity all kinds of things become possible.

Far right: The Skylab 4 crew are no doubt glad that they are safely way above this massive spiraling storm center, which back on Earth is the destructive hurricane Ava.

Below: This still from a 16-mm movie taken on board Skylab lacks crispness technically but offers an interesting view of the space station. It shows Edward Gibson, who has just emerged from the EVA hatch. Note the umbilical snaking toward the photographer shooting the film, Gerald Carr.

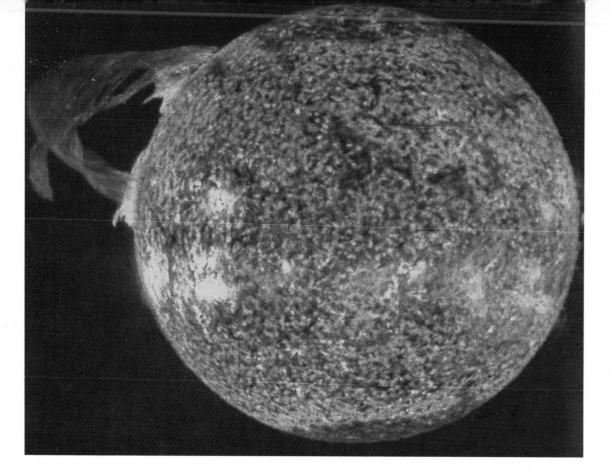

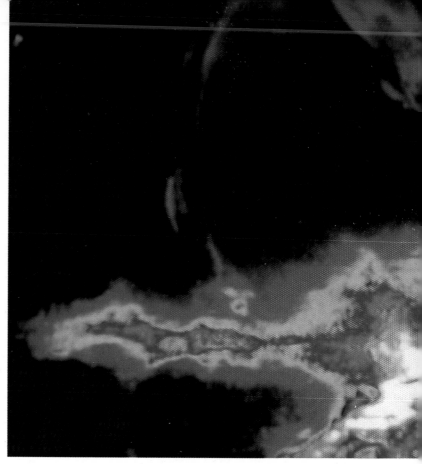

its 10 months of observations, acquired more data about the Sun than Earthbound astronomers had done in centuries. This convincingly demonstrated the quantum leap in the quality and quantity of astronomical observations that was possible with space-based instruments.

Photographic observations of the Earth in ordinary light and at selected wavelengths, via the so-called Earth-resources experiments package (EREP), gathered data useful in many disciplines, from agriculture and geography to meteorology and oceanography. In materials-science the astronauts conducted experiments to make new alloys and grow purer crystals. They found that in zero-gravity such materials could be fabricated relatively easily.

In the field of natural science, the astronauts on Skylab 3 looked into the problems spiders would face in space. This was one of the student experiments Skylab carried. Could space spiders spin webs well enough to catch space flies? Unaccustomed as they were to spinning in zero-g, two spiders called Arabella and Anita spun webs with only slightly less expertise than they did on Earth.

Skylab's fiery return

After the Skylab 4 crew vacated the space station early in 1974, it was mothballed. It was not in A1 condition, but NASA were hopeful that a few years later another crew would make a visit in the space shuttle, then under development. However, two things worked against this. To begin with the vestiges of atmosphere, even at Skylab's original height (270 miles, 435 km), gradually slowed it down. This meant that it descended into a lower and lower orbit as time went by. NASA could do nothing about this because Skylab had no engines of its own, apart from low-powered maneuvering thrusters.

By 1979 Skylab was orbiting perilously close to the atmosphere. The space shuttle, however, was not ready and could not be used, as had once been hoped, to drive Skylab into a higher orbit. There was no doubt about it, the 77-ton (70-tonne) space station was coming back to Earth. The Earth's population was more than a little worried about its arrival. As Skylab descended still further, NASA calculated that it would re-enter the Earth's atmosphere in July, within days of the tenth anniversary of Apollo 11's lift-off to the Moon.

As the fateful day approached, 'Skylab watch' parties were held throughout the US, and Skylab survival kits including a crash helmet and a target did a roaring trade! Millions of radio listeners were urged to use their combined psychic power to lift Skylab into a higher orbit. It didn't work. Skylab just kept coming.

NASA calculated that when Skylab did start to break up it would scatter debris in a 'footprint' 4000 miles (6500 km) long and 100 miles (160 km) wide. They also calculated the probability of a piece striking a human being as 152:1. To many, these were not good odds. Only people living north of the 50° North and south of the 50° South latitudes were completely safe and outside Skylab's orbital path.

Skylab's last days in orbit were tracked meticulously by the worldwide network of NASA and NORAD (North American Aerospace Defense Command) space tracking stations. It was going to come down on 11 July, on its 34,981st orbit. But it looked as if it was going to start breaking up over Canada, endangering the Montreal region and part of Maine. To try to delay re-entry, mission control at Houston ordered Skylab's attitude jets to fire to start it tumbling.

Thanks to this final maneuver, Skylab did not hit the atmosphere

OXYGEN
TANK

Left: When Skylab finally falls out of the skies in July 1979, it breaks up over the Pacific and Western Australia. Some very large pieces survive its fiery descent, including this oxygen storage tank.

Below: The US/USSR Apollo-Soyuz Test Project (ASTP) captures the world's imagination when it is announced. Cooperation, not competition between the United States and the Soviet Union is perhaps the way space exploration should progress. The prime crews for the mission, which is scheduled for summer 1975, are Thomas Stafford, Donald Slayton and Vance Brand (US); and Alexei Leonov (standing) and Valery Kubasov (USSR).

STAFFORD

NASA

SLAYTON

NASA

BRAND

NASA

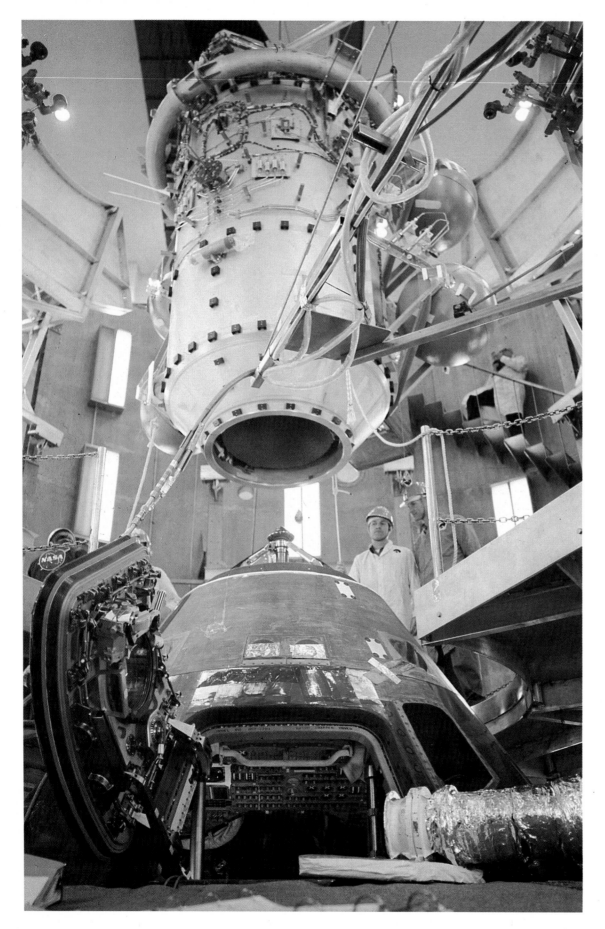

until it was over Ascension Island in the South Atlantic. Unfortunately, this meant that it finally broke up over Western Australia. Perth saw a spectacular fireworks display as the disintegrating Skylab streaked overhead. The biggest sections came down beyond Kalgoorlie, several hundred miles east of Perth in a fortunately sparsely populated region.

Comrades in space

The final chapter in Apollo's history, the Apollo-Soyuz Test Project, had its origins in tentative contacts between American and Russian space scientists in the early 1960s. The first official agreement on space cooperation was signed by NASA Deputy Administrator Hugh Dryden and Anatoly Blagonravov of the Soviet Academy of Sciences in 1962. It provided for the exchange of data from scientific and meteorological satellites. A 1965 agreement led to the exchange of information on space medicine.

Some five years later NASA suggested making cooperation between the US and the USSR more tangible by means of joint dockings in space. Regular meetings then started, alternately in Russia and the United States, to discuss compatible rendezvous and docking systems. They culminated in the agreement signed by President Nixon and Premier Kosygin on 24 May 1972, which approved among other things a joint Apollo-Soyuz space flight. The Apollo-Soyuz Test Project (ASTP) was born.

Immediately preparations got underway. Prime and back-up crews were chosen and training began. The American crew of Apollo were Thomas Stafford (commander), Donald Slayton and Vance Brand. Stafford was considered a leading expert on the rendezvous and docking maneuvers that would be required in the joint flight. He had been in space three times, twice on Gemini flights (6 and 9) and on Apollo 10, the dress rehearsal for the first Moon landing. Slayton was one of the 'original seven' Mercury astronauts selected way back in 1959, but had never flown before because of a suspected heart condition. When he went into orbit, he would, at the age of 51, become the oldest man to venture into space. Vance Brand too was a rookie.

The Russian Soyuz had a two-man crew. Commander was Alexei Leonov, who had pioneered spacewalking in orbit from Voshkod 2 in March 1965. Traveling with him would be Valery Kubasov, who had previously flown on the Soyuz 6 mission in 1969.

Communications problems

The two crews, and back-ups, exchanged visits to their respective training centers for joint flight training. The American astronauts flew to the Gagarin Cosmonaut Training Center at Zvezdnyy Gorodok near Moscow, better known as Star City. The Russian cosmonauts visited the Johnson Space Center at Houston for training and flight simulations. They also toured the facilities at the Kennedy launch site.

Mission controllers at Houston also exchanged visits with their

Left: The docking module that will link the Apollo (US) and Soyuz (USSR) spacecraft in orbit will be carried into space already mated with Apollo.

Here, fit checks between the module and Apollo are being carried out at Kennedy early in 1975.

counterparts at Kaliningrad, where Russia's mission control is located. An elaborate communications network had to be established between the two centers to allow continuous contact during the mission. To extend communications between the spacecraft and the control centers, NASA utilized its powerful communications satellite ATS-6. In stationary orbit 22,300 miles (35,900 km) above the equator, ATS-6 could provide communications for 50 per cent of the time. It would be the first time that communications to manned spacecraft would be routed via an orbiting satellite; something, however, that is now done routinely on shuttle flights by the TDRSs (tracking and data relay satellites).

As to be expected, one of the biggest communications problems was the language barrier. The success and safety of the mission would hinge on the swift and unambiguous exchange of information between control centers on the ground and the two crews in orbit. For the participating astronauts and cosmonauts it was vital, so the crews learned each other's language. When communicating to each other during the joint phase of the flight, the astronauts would speak Russian and the cosmonauts would speak English. A large volume containing detailed operational instructions for the mission in Russian and English was carried on each spacecraft as a safeguard.

Working groups, meeting alternately in the US and Russia, met 44 times over the three-year period before the flight was due to lift-off. Their work encompassed communications, life-support, tracking and guidance, overall mission planning and coordination as well as modifications to the two spacecraft to allow rendezvous and docking.

The two craft were to be linked by a chamber that had docking ports at each end. This docking module would also act as an airlock, required because the atmospheres in Apollo and Soyuz were different. Whereas Soyuz normally had a nitrogen/oxygen mixture at sea-level pressure, Apollo used pure oxygen at only a third of this pressure. The docking module would be carried into orbit mated to the Apollo CSM.

Handshake in orbit

Training and launch preparations began to accelerate in both countries at the beginning of 1975, as they aimed for a 15 July launch. The US launch vehicle, a Saturn IB with the Apollo CSM and mated docking module on top, rolled out to the launch pad of Complex 39 in March. The final countdown began on 11 July. The same day, far away on the Russian steppes at the Baikonur launch site, the Soyuz spacecraft was moved to its pad.

Launch day arrived, and it was Soyuz that lifted off first, precisely on time at 8.20 am EDT. The launch was shown live on TV, the first time this had ever been done. US Ambassador to Russia, Walter Strossel, watched the launch from Baikonur. At the Kennedy Space Center Soviet Ambassador to the US, Anatoly Dobrynin, watched Apollo lift off seven and a half hours later, precisely on time at 3.50 pm EDT. Over the next two days the two orbiting craft, launched into separate orbits, fired their engines in a series of burns that brought them into docking formation.

At 12.09 pm EDT on 17 July they docked together over Europe, six minutes ahead of schedule. It was Stafford who conducted the perfect docking maneuver.

Above: While the flight hardware is being readied at Kennedy, the ASTP crew are in training at the Johnson Space Center in Houston. Astronaut Vance Brand (left) is pictured here with cosmonaut Valery Kubasov in a mock-up of Soyuz's orbital module.

Right: On 17 July 1975 the Apollo crew have this view of Soyuz as the two craft slowly inch toward one another.

Stafford and Slayton then transferred to the docking module and pressurized it to about two-thirds sea-level pressure. The cosmonauts in Soyuz reduced their pressure to the same level. Three hours after docking Stafford opened the hatch of the docking module connecting with Soyuz. 'Come in here,' he said in Russian to Leonov who came floating through the hatchway. 'Glad to see you,' Leonov replied in English. At 3.19 pm EDT they shook hands, cementing their personal friendship and symbolizing a new era of cooperation rather than competition between the world's two great space powers.

The two crews of the first international flight exchanged flags, medallions and plaques. They also autographed historic early publications of Russia's father of space travel, Konstantin Tsiolkovsky, and the American rocket pioneer Robert Goddard. They visited each other's spacecraft and conducted a variety of experiments together. They shared each other's food, the American astronauts being introduced to borscht, jellied turkey and black bread.

The end of the beginning

The Apollo and Soyuz spacecraft stayed docked together until 19 July. After an early morning separation and re-docking maneuver,

Right: On the historic rendezvous day both crews sign a certificate inside the Soyuz orbital module. Writing in the picture is Valery Kubasov.

Left: A hard dock between Apollo and Soyuz is successfully achieved. This artist's impression shows how the two craft appear in orbit.

the craft began to make their separate ways back to Earth. Soyuz made an uneventful re-entry and touched down on 21 July. Apollo's return three days later was far from uneventful. Just before splashdown in the Pacific, the crew noticed yellowish fumes in the command module which set them coughing and irritated their eyes.

On splashdown the command module flipped over, leaving the astronauts hanging upside-down from their couches. With choking fumes still coming into the cabin, Stafford unstrapped himself and managed to reach the oxygen masks stored on board. He and Slayton donned their masks, but had to help Brand, who by now had become unconscious. He came round after about a minute. Then the capsule righted itself and the hatch could be opened to admit fresh air.

The crew recovered quickly but were hospitalized in Honolulu for several days before being given a clear bill of health.

It turned out that the yellowish gas in the capsule had been nitrogen tetroxide from the maneuvering thrusters. It is a toxic gas that attacks the lungs and could have been fatal in large quantities. The gas had entered the capsule through an air vent after the thruster units had been accidentally left on after parachute deployment.

With that potentially disastrous splashdown, on 24 July 1975, the Apollo missions finally came to an end. In seven years Apollo spacecraft had carried 45 astronauts on 15 flights in Earth orbit and nine visits to the Moon. It was the end of the expendable era in American manned space flights. Henceforth American astronauts would be ferried into orbit in a re-usable craft, a space-liner.

Left: Feeling peckish, astronauts Thomas Stafford and Donald Slayton take advantage of Russian hospitality and sample their host's food, from toothpaste-type containers. Note the Vodka labels on the tubes!

Chapter 5

A Magnificent Flying Machine

Until an American foot had been planted on the lunar surface NASA was too preoccupied to put forward any concrete proposals about what direction space flight should take after Apollo. In September 1969, however, two months' after Neil Armstrong's historic step, NASA submitted a report to a Space Task Group set up by President Nixon outlining its suggestions for the future. It recommended that there should be a re-usable space transportation system that could shuttle back and forth between Earth and space.

This was by no means a novel idea. In the 1950s Wernher von Braun pointed out the economic benefits of re-usable Earth-to-orbit cargo carriers. The world's two foremost space societies, the American Rocket Society and the British Interplanetary Society, came up with ideas for rocket airplanes and re-usable manned orbital rocket systems in the same decade.

But in the pre-space flight era these concepts were not technically feasible. In any case, the military need to develop long-range ballistic missiles kept the emphasis on conventional expendable rocketry. But as space flight gathered momentum in the 1960s, the need for a more economical launch system became more pressing. It also became feasible as a result of the advances that had been taking place in aerospace technology; with the spectacular flights of the X-15 rocket plane; the lifting-body research craft (such as the X-24B, the last of this family of wingless hypersonic gliders); as well as the Mercury, Gemini and Apollo spacecraft.

The space plane options

Following the Space Task Group's recommendations, President Nixon announced approval of a re-usable space transportation system in March 1970 and set in motion two years of feasibility and engineering studies for a space plane. In the initial thinking it was to be a fully re-usable two-stage vehicle that would launch vertically and land horizontally. The first stage would boost the second stage well above the atmosphere and then return to base. The second stage would then fire and thrust itself into orbit. Both stages would be manned by a crew of two, and the second stage would also be capable of carrying a number of passengers.

Considerations of cost and safety forced the designers to

These pages: On its third flight into space, orbiter Challenger blasts off spectacularly at 2.30 am local time on 30 August 1983. Night turns into day as it leaves the launch pad for a highly successful six-day mission. It will also land at night just as dramatically.

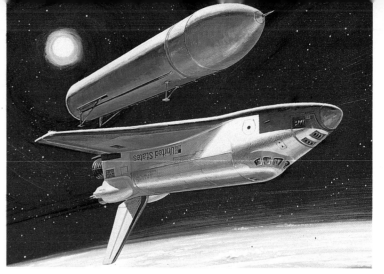

Left: The space shuttle orbiter rides piggy-back into space atop a huge external fuel tank. Attached to the sides of the tank are twin solid rocket boosters. The orbiter's main engines and the boosters all fire together at lift-off to blast the shuttle into space. The boosters separate when their fuel runs out after about 2 minutes, and parachute back to Earth.

Above: Some six minutes after booster separation, the external tank runs out of fuel, and is jettisoned in turn.

Below: At the end of the mission the orbiter re-enters the atmosphere traveling at about Mach 25 (25 times the speed of sound). The atmosphere acts like a brake and slows down the orbiter to a safe landing speed.

abandon the idea of a piloted first-stage booster. It would be unmanned, but recoverable. It could have either liquid-fuel or solid-fuel engines. The difficulty of recovering liquid engines undamaged and their much greater initial cost, however, eventually ruled them out.

Economics also dictated changes in the design of the upper-stage vehicle as time went by. The first idea that it should be a completely self-contained vehicle, carrying its own fuel and oxygen supplies, had to be abandoned. It evolved into a much smaller craft that carried the fuel and oxygen necessary for launch, in a separate tank. When the tank was empty, as the craft entered orbit, it would be jettisoned.

So, by the end of 1971, the following design for the launch vehicle of the future was recommended. There would be a fully re-usable manned orbiting craft, which would take its propellants from an expendable external tank. The orbiter/fuel-tank would be launched with solid-propellant boosters, which would later be recovered and re-used. In January 1972 the President approved the plan, recommending that: 'The United States should proceed at once with the development of an entirely new type of space transportation system designed to help transform the space frontier of the 1970s into familiar territory.... It will revolutionize transportation into near space by routinizing it.... It will take the astronomical costs out of astronautics'.

The space shuttle as we know it today was born.

Prototype *Enterprise*

Rockwell International of Downey, California, won the contract for building the major piece of hardware in the shuttle system, the manned orbiting craft, or orbiter. Some 122 feet (37 meters) long, the orbiter has a delta-wing with a span of 78 feet (24 meters). It is much the same size as a medium-range airliner like the DC-9. But it has rocket motors in its tail, not jet engines, and these motors are used only for the journey into space.

When the orbiter returns to Earth, it is as a glider – the heaviest glider ever, tipping the scales at some 98 tons (88 tonnes). Before it hits the atmosphere, though, it must withstand the searing heat of re-entry, generated by friction with the atmosphere. To protect the crew inside, the outside of the orbiter is covered in heat-resistant tiles and other insulation.

The first orbiter to be built, however, had neither motors nor insulation. It was a prototype, constructed solely for test purposes. At the direction of President Gerald Ford, this trail-blazing orbiter was named *Enterprise*, after the famous starship in the Star Trek television series. Alas, shuttle orbiter *Enterprise* was destined never to fly into orbital space, let alone travel at warp speeds through intergalactic space!

Enterprise did, though, fly in the atmosphere in a series of approach and landing tests in 1977 to evaluate the aerodynamic behavior of the orbiter design. These tests took place at the Dryden Flight Facility at Edwards Air Force Base in California. The first flights were in February, with *Enterprise* being carried aloft on a specially converted Boeing 747. They were unmanned, and the orbiter remained fixed on the top of the 747 throughout the flights.

Below: A detailed cutaway of the space shuttle orbiter, the first true aerospace vehicle. Its airframe, surprisingly, follows conventional airplane construction practice. It is made of high-strength aluminum alloys, but is covered with highly efficient insulation which forms a barrier against the searing heat of re-entry.

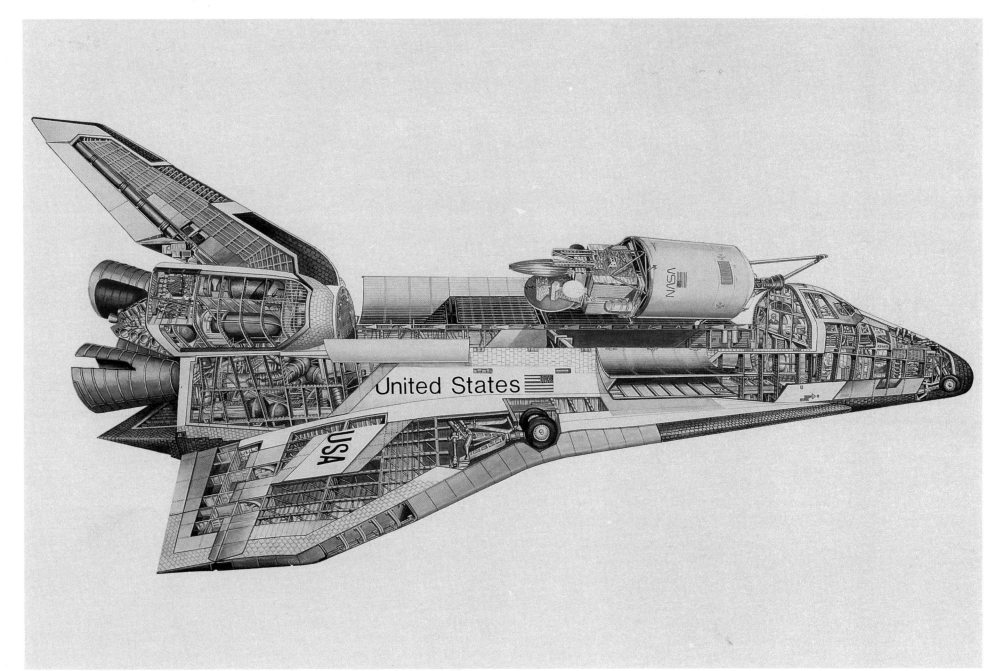

In August a two-man crew made the first free-flight in *Enterprise*. It cut loose from the 747 at some 23,000 feet (7,000 meters) and glided immaculately back to the ground, making a perfect runway landing. Further free-flights confirmed that the aerodynamics of the orbiter in low-speed flight were exactly as predicted from wind-tunnel tests. From the pilots' viewpoint, *Enterprise* handled well, almost, they said, like a fighter.

Following the successful approach and landing tests, *Enterprise* was flown to the Marshall Space Flight Center in Alabama, where it was subjected to rigorous vibration testing. Again it passed with flying colors, proving itself structurally sound. The vehicle then transferred to the Kennedy Space Center, where it would serve as a test-bed for verifying procedures for orbiter handling and mating with the other shuttle components – the solid rocket boosters and the external tank.

'The Flying Brickyard'

Enterprise arrived at Kennedy a little over two weeks after the first operational orbiter, *Columbia*, had been transported from Downey (on 24 March 1979). Like *Enterprise*, *Columbia* had flown in on the 747 carrier jet. It had looked a sorry sight, for thousands of the tiles of its heat shield were missing. Many others had been damaged during flight and subsequently had to be replaced.

The tiles, in fact, gave the manufacturers of the shuttle one of their biggest headaches. Made of silica fibers, the tiles had brilliant heat-dissipating properties, but were fragile. They also proved much more difficult to bond to the orbiter's airframe than expected. In all, over 30,000 tiles had to be stuck on *Columbia*, some as thick as bricks. The orbiter was well nicknamed 'the Flying Brickyard'.

Development and bonding problems with the tiles were major factors in the delay in introducing the space shuttle into NASA's

Left: Construction proceeds on the first prototype orbiter Enterprise *at Rockwell International's manufacturing facility in California.*

Above: In 1977 Enterprise *is hauled the 36 miles (58 km) from Palmdale to the Dryden Flight Facility along the ordinary highway, enabling the public to see this revolutionary flying machine for the first time.*

Right: More than 22,000 feet (6700 meters) above the Mojave Desert in California, Enterprise *separates from a specially converted Boeing 747 carrier jet on its first free flight in August 1977. An aerodynamic fairing is in place in the tail, over the engine pod.*

launch stable. On the original schedules, *Columbia* should have been flying by the end of 1979, instead of which it was still on the ground being tiled.

At the same time, problems with the shuttle main engines also bit into the schedule. The engine design was completely new and demanded new technologies. It was required to work at unprecedented pressures (up to 200 times atmospheric pressure) and to be capable of throttling and multiple starts. Hitherto, launch rocket engines had been used only once, and had a lifetime measured in minutes. The space shuttle main engines, on the other hand, were required to last for over seven hours, and for at least 50 missions. Eventually the problems were overcome, resulting in the most efficient rocket engine ever produced. Though it weighs less than 7000 pounds (3000 kg), it has the power output of seven Hoover dams!

By November 1980 the problems with tiles and engines seemed at last to have been overcome, and *Columbia* was moved the few yards from the orbiter processing facility into the Vehicle Assembly Building (VAB). There it was hoisted into position and mated with the external tank, already assembled with the solid rocket boosters (SRBs) on the mobile launching platform. In the early morning of 29 December came the historic first rollout of the new generation launch vehicle that was destined to change the face and pace of space exploration for the rest of the century. *Columbia*, proceeding at the sedate pace of 1 mph (1.6 km/h), moved towards launch pad 39A. By the evening it was secured and, bathed in floodlights, stood poised for its first journey into space.

The launch date had been set for the end of March 1981. The

Above: A beautiful shot of Enterprise *on its first free flight, gliding under pefect control towards the runway at Dryden Flight Facility, Edwards Air Force Base.*

Left: Enterprise *finishes its final approach and landing test at Dryden in October 1977 after a perfect 2-minute free flight. Note that the tail fairing has been removed to assess the orbiter's performance under mission conditions. With or without the fairing, the orbiter performs like a champ.*

Right: Now a fully accredited flying machine, Enterprise *stands on the runway apron at Dryden. In the foreground are the men who piloted it on the approach and landing tests. From left to right they are Gordon Fullerton, Fred Haise, Joe Engle and Richard Truly.*

Above: The shuttle main engines are the most efficient engines ever made. Burning a mixture of liquid hydrogen and liquid oxygen, they develop a thrust of up to 450,000 pounds (205,000 kg) for a weight of only 6700 pounds (3000 kg). Here an engine is being test-fired at NASA's National Scientific Technology Laboratory (NSTL) at Bay St. Louis in Mississippi.

Right: At the Thiokol/Wasatch test site in Utah, a solid rocket booster (SRB) for the shuttle is test-fired. Measuring some 149 feet (45.5 meters), the SRB develops a thrust of more than 2.6 million pounds (1.2 million kg). The propellant is a mixture of aluminum powder (fuel) and ammonium perchlorate (oxidizer) in a synthetic rubber binder.

crew on this first mission, designated STS (space transportation system)-1, were to be veteran astronaut John Young, with two Gemini and two Apollo missions behind him, and rookie Robert Crippen. In all previous manned space programs spacecraft had first been flown into space unmanned before manned flights were attempted. The shuttle was different. Though each item of shuttle hardware had been separately ground tested, they had not been tested together nor in a live launch. Young and Crippen would indeed be test pilots of the old school, flying an unproven machine both into space and back. They had spent hundreds of hours 'flying' the orbiter in the realistic simulators at the Johnson Space Center. Simulators, however, are only as good as the data fed into them. There was no real-time data relating to the launch of a delta-winged aerospace craft or its return through the atmosphere at speeds up to 25 times the speed of sound. When it eventually took off, *Columbia* would be flying into the unknown.

Countdown

As soon as *Columbia* was installed on the pad, technicians descended upon it to ready it for its pioneering mission. By February 1981 preparations were advanced enough to permit a practice countdown and test firing of the troublesome engines. On the 20th the engines roared into life for a 20-second burn, and shut down as planned. One of the last hurdles had been cleared. Later a full dress rehearsal for the launch team and crew was held. With the success of this, the launch date was set for 6.50 am Friday, 10 April.

As the week of the launch attempt went by, thousands of reporters and photographers from the world's press descended on

Left: Bonding the tiles to the orbiter is a tedious and time-consuming job. Each tile is unique and cannot be interchanged with any other tile. It is identified by a computer number. It is bonded to a Nomex felt pad, which is in turn bonded to the airframe. Columbia is covered with more than 33,000 of these tiles. Small wonder it is called the 'flying brickyard'!

Top right: While the hardware for the space shuttle is being developed, NASA begins recruiting more astronauts — both pilots and mission specialists — who will be responsible for payload operations and general mission duties. Among the astronaut candidates are the first women considered for space flight in America. Here they pose with a spacesuited mannequin. From left to right they are Rhea Seddon, Kathryn Sullivan, Judith Resnik, Sally Ride, Anna Fisher and Shannon Lucid.

Right: Columbia rides piggy back on its 747 carrier jet as it leaves Dryden Flight Facility in California on 3 March 1979, bound for what will become its home base, the Kennedy Space Center in Florida. Note that the tile heat shield is not yet complete.

Kennedy Center. With three days to go, the final countdown began. Crowds began to pour into Cocoa Beach and Titusville, with just one thing in mind – to see the historic first flight of *Columbia* that would introduce a new era in space travel. By Thursday evening tents and campers were pitched all along the roads within the Kennedy Space Center and the shoreline of the Indian River. Not since the Apollo flights had such crowds gathered.

The countdown proceeded with remarkable smoothness. As dawn broke, all eyes were riveted on the distant launch pad that had been bathed all night in powerful floodlights. In *Columbia* now were Young and Crippen, suited up, strapped in, and raring to go. Then the countdown went into a scheduled hold with only nine minutes to go before lift-off, at T -9. Then inexplicably the computers that would mastermind the flight began to malfunction.

There are, on board the orbiter, five computers – four main and a back-up. The back-up computer continuously reviews data processed by the four main computers. If it registers differences in the data from either of the other four, it issues warning signals.

On that Friday launch day at T -9, *Columbia*'s computers suddenly developed a glitch. They began to work out of phase – only by milliseconds, but in computer operation that is a long enough time to create problems. A NASA official summed it up by saying that the computers weren't talking to each other properly. They were answering questions, as it were, before they were asked. Mission control director Neil Hutchinson labeled it 'a very statistically improbable circumstance'.

Improbable though it may have been, it fouled up the launch. Attempts for a 10.20 am lift-off after reprogramming of the computers were abandoned when the trouble persisted. Over three-quarters of a million tired and disappointed spectators went away from the space center that morning. By the evening, however, the bug in the computer system had been isolated, and the problem was solvable. On Saturday the countdown resumed for a Sunday morning lift-off at 7 am.

'Go, go, go!'

The crowds returned on Saturday and, with the launch team and press corps, prepared for the second night in three without sleep. This time, however, they were not to be disappointed. The countdown went into — and this time came out of — the scheduled ten-minute hold at T -9.

The countdown clock ticked down to zero. A few milliseconds after 7 am *Columbia*'s three main engines roared into life. Clouds of steam rose around the launch pad as water cascaded over it to reduce the reflected energy. Then the twin solid-rocket boosters

Above: Columbia *being towed towards the orbiter processing facility at Kennedy after flying in from Dryden. It arrives in less than perfect condition. Some dummy tiles stuck on the orbiter for the flight fell off and damaged many of the real tiles. A lot of work remains to be done on* Columbia *before it can attempt its first lift-off.*

Left: On 1 May 1979 a full shuttle stack rolls out of the Vehicle Assembly Building (VAB) for the first time. It is mounted on the mobile launch platform and carried by the huge eight-crawler tracked transporter. This is a dummy run, with Enterprise *mated with the external tank and solid rocket boosters.*

Right: The Enterprise *shuttle stack is maneuvered into position on Complex 39, launch pad A. This is an excellent view of the pad area, showing very well the rotary service structure, which swings around to give access to the orbiter fuselage and payload bay. Note the scale of the hardware from the size of the vehicles around the pad.*

blazed into life like firecrackers. As thrust built up, the clamp-down bolts that held the shuttle down to the launch platform were severed. *Columbia* leapt into the air on a pillar of flame and smoke. 'Go, go, go' came the cries as it was urged on its way by the yells and cheers of the hundreds of thousands of spectators.

Within seconds *Columbia* cleared the launch tower and was rolling into a 'heads-down' position (with the boosters and fuel tank on top). Within a minute it had virtually disappeared from sight, hidden by the murky exhaust from the boosters. Two minutes 12 seconds into the flight and 30 miles (45 km) high, the boosters ran out of fuel and separated, arcing away and later parachuting back to an ocean splashdown. The main engines continued firing. '*Columbia* is now committed for space flight', reported mission control, 'Young and Crippen can no longer return to launch site'.

Six minutes later the main engines cut off, and *Columbia* jettisoned the huge external tank. This arced up and over and then broke up as it had a blazing re-entry into the atmosphere. Young and Crippen then allowed *Columbia* to coast for a while before firing the two engines of the orbital maneuvering system (OMS) to take them finally into orbit.

Far left: It is 24 November 1980 and here inside the Vehicle Assembly Building Columbia *is waiting to be lifted up to mate with the external tank. This will complete the assembly of the shuttle stack for the first mission of the space transportation system (STS-1).*

Left: On 29 December 1980 the huge doors of the VAB open, and Columbia *starts its 6½-hour journey to the launch pad. There, final preparations will be made for the shuttle's maiden flight in the spring.*

Above right: Crippen (seated) and Young are pictured here inside the cockpit of Columbia. *They have spent hundreds of hours practicing flying the orbiter into space on the shuttle mission simulator at the Johnson Space Center, Houston, in conjunction with mission control there. Note the bewildering conglomeration of switches, gauges, instruments, keyboards and controls.*

Right: Astronaut John Young (left), veteran of Gemini and Apollo missions, will be commander of the first shuttle flight (STS-1). Pilot will be rookie Robert Crippen. They are pictured dressed in their flight suits, which are based upon test-pilot suits. It will be very much as test pilots of the old school that they will be flying the shuttle, the most complex flying machine yet devised.

The long-delayed shuttle, which had been disparagingly labelled by some the 'Space Lemon' and the 'Spruce Goose of Outer Space' (referring to Howard Hughes's ill-fated monster airplane), had passed its first test with flying colors. Commented Young, whose heart beat had remained at a steady 85 beats per minute during lift-off: 'This vehicle is performing like a champ. I've got a super spaceship under me.' Said Crippen, whose heart beat had peaked at 135: 'What a feeling! What a view!'

'Right on the money'
Up in orbit at an altitude of 170 miles (275 km), *Columbia* was looking good. Only slightly worrying was the loss of some tiles on the engine pod. These were in a non-critical area as far as re-entry was concerned. But had any more tiles been lost in critical areas like the underside and nose? There was no way of knowing. Otherwise there were no problems, and *Columbia* was cleared for a full 36 orbits as planned.

The 36th orbit found the crew preparing for their de-orbit burn. They turned round *Columbia* until it was traveling tail first. Then they fired the OMS engines again. The orbiter slowed and began to drop towards the Earth's atmosphere, heading for a runway landing at Edwards Air Force Base (AFB) in California. For the first time a winged craft would be attempting re-entry. Could it, would it, survive the massive deceleration and 1500°C (2700°F) temperatures caused by air friction?

It impacted the atmosphere belly-on, and the tiles on the underside and nose glowed red-hot. Communications ceased between mission control and the crew as the ionized air around the orbiter prevented the passage of radio signals. *Columbia* was in the communications blackout for 21 minutes. Then communications were re-established. All was well. Commented Crippen: 'What a way to come to California!'

As the hefty space glider got nearer and nearer Edwards, it did a series of S-turns to reduce speed, then it swooped steeply towards the runway. 'You're really looking good. You're right on the money,' said mission control, as *Columbia* made its final approach. A scant nineteen seconds before touchdown the undercarriage came down. The two rear wheels hit first, making little puffs of dust on the long desert runway. *Columbia* was back after a near-flawless 54-hour maiden voyage, during which it had covered a little under a million miles (1.6 million km).

'Do it again, *Columbia*'
Just two weeks after its pinpoint landing at Edwards, *Columbia* was back home at Kennedy Space Center and ready for checkout for its next journey into space, planned for the following fall. The NASA troubleshooters moved in to explore and rectify the 'anomalies' shown up on the first flight. First there was a tile problem. Sixteen had been partly or completely lost on lift-off. Nearly 150 others were damaged, probably by ice breaking off the

Above: A month before scheduled lift-off, Young and Crippen take part in a practice on-the-pad escape from the launch tower which would save their lives in an emergency.

Right: During March 1981 the shuttle launch pad is ablaze with light and buzzing with activity. Lift-off day is fast approaching. Soon the final countdown will begin.

Inset right: At the Kennedy site near the Vehicle Assembly Building, several thousand photographers and reporters of the world's press corps have spent a sleepless night counting down the hours and minutes with shuttle control.

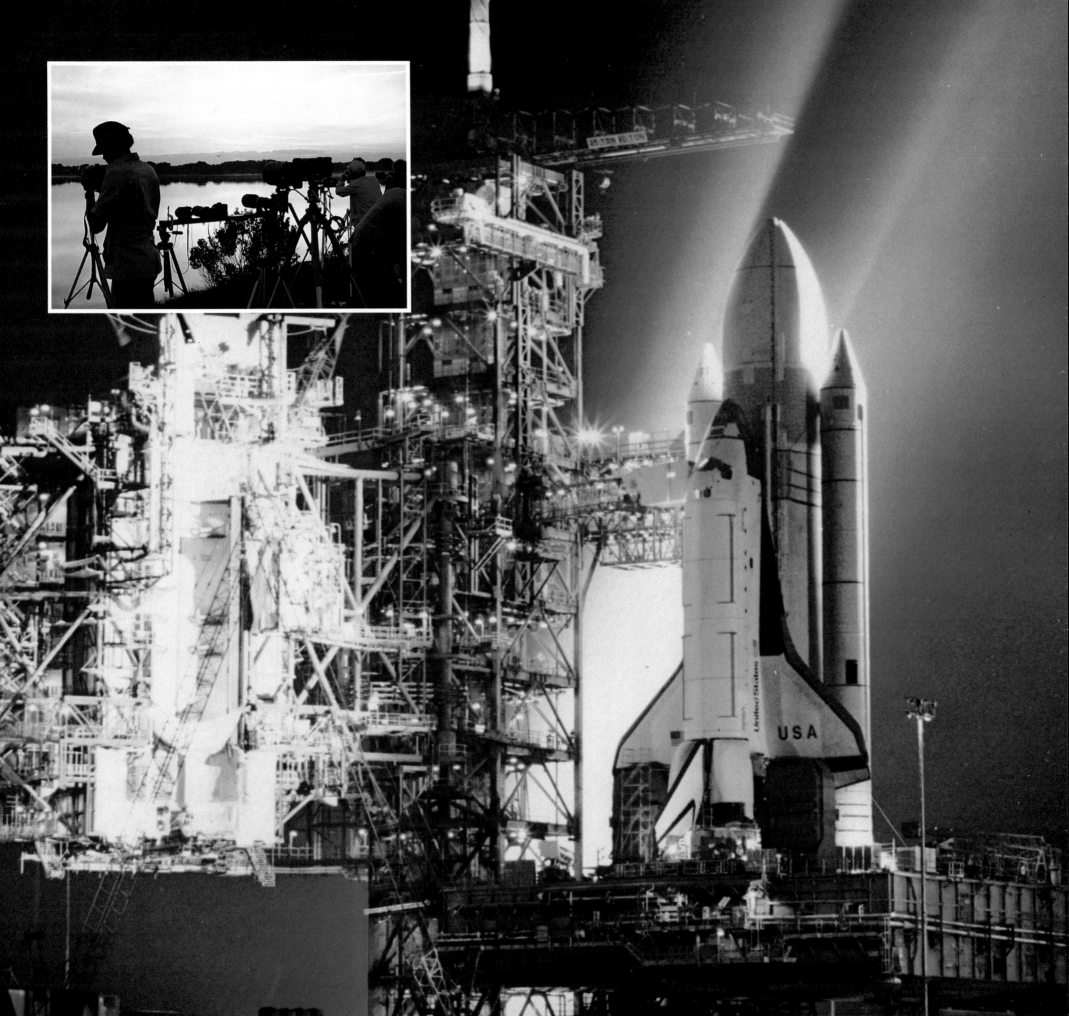

Left: Along NASA Causeway, the road that leads into the Kennedy Space Center, people have been gathering in their thousands. Not since the Apollo days has the Center seen such crowds or experienced such excitement.

Right: The countdown clock ticks away, and the assembled photographers check their focus, adjust apertures and speeds as the Sun rises on a fine day.

Below: Then the countdown clock ticks down to zero. From the press site you can see a cloud of steam rise from the launch pad. Columbia's main engines ignite. The tension is unbearable. Then the thunder hits your ears, and the shuttle leaps from the pad. We have lift-off!

supercold external tank that contained the liquid hydrogen and oxygen propellants.

The intensity of the reflected sound energy coming from the flame tubes beneath the booster rockets was far higher than expected. It had damaged the external tank struts and may well have been responsible for shaking loose some of the tiles. Inside the orbiter the most significant problem was the failure of its zero-g toilet, which fortunately happened only a few hours before touchdown.

So during the summer, tiles were replaced, the toilet was modified, flame-tube water sprays were installed, and by September *Columbia* was back on the launch pad heading for an October lift-off. Then the gremlins struck again. Nitrogen tetroxide oxidizer being pumped into the orbiter's reaction control system tanks spilled onto the tiles on the nose and dissolved the bonding cement beneath them. More than 350 tiles had to be removed and rebonded.

The STS-2 launch was re-scheduled for 4 November 1981. Early that day astronauts Joe Engle and Richard Truly climbed into *Columbia* and counted down with launch and mission control until T-9 minutes. Then things started going wrong. Low pressures were detected in oxygen tanks in the external tank and orbiter. They did not appear serious and the countdown was resumed. But at T-31 seconds, the computers decided they didn't like the readings and called a halt. Then the auxiliary power units in the orbiter began playing up. The launch had to be scrubbed.

On 14 November, however, *Columbia* did make it off the launch

Left: Columbia *leaves the pad on a pillar of flame and smoke, produced mainly by the prodigiously powerful solid rocket boosters. Never before have solid rockets been used for a manned space flight. Will they prove reliable? Indeed will everything prove reliable on this untested vehicle? This picture is taken by a specially protected camera relatively close to the pad and automatically triggered.*

Right: This sequence of pictures shows the historic first lift-off as viewed from the press site. Within seconds the shuttle is high in the sky and accelerating fast. In less than a minute it is lost from view atop the contrail of smoke.

Right: Some 28 miles (45 km) above the Atlantic Ocean, Columbia *sheds its solid rocket boosters. Its own engines, fueled from the external tank, continue burning, thrusting it ever faster, ever higher. This picture was taken by a long range tracking camera.*

pad. For the first time a craft was climbing into space for a second mission. But within two hours of lift-off alarms started flashing in the cockpit. One of the orbiter's three fuel cells, which produce electricity and drinking water for the crew, failed. The mission was curtailed to a two-day 36-orbit trip like the first, instead of the planned five days.

The crew nevertheless managed to fulfill some mission objectives. In particular they put the Canadian-built remote manipulator arm throught its paces. This 50 foot (15 meter) appendage is the shuttle's crane, designed to lift satellites out of the payload bay and also retrieve them from orbit. The arm worked perfectly, as did the experimental shuttle imaging radar, which produced images of the Earth's surface using radar techniques. This had never been done before from orbit.

Last of the R and Ds

After STS-2, two more flights were planned to complete the scheduled research and development program for the shuttle. If they went well, then the spacecraft would become operational and start carrying commercial payloads into orbit, for cash.

The cliff-hanging countdown delays that had plagued the first two shuttle missions were virtually absent on STS-3. *Columbia* punched its way into the hazy blue Florida skies only one hour behind schedule on 22 March 1982. It was riding this time on an external tank with a new look. The tank this time around was orange-brown rather than white, lacking the gleaming white paint

of the previous tanks. This practice has continued ever since, saving an incredible 600 pounds (275 kg) in weight and $15,000. Every little helps.

Flying *Columbia* this time were Jack Lousma and Charles Fullerton. Lousma was making his second space flight, his first having been in Skylab in 1973. Then he had suffered a bad attack of space sickness, and unfortunately went down with another this time. The orbiter, too, suffered a recurrence of an old ailment – tile-shedding at lift-off. Some 36 were lost and 20 damaged. Later, the toilet clogged again, two TV cameras stopped working, a teleprinter went wild and, more seriously, three communications links with ground control failed.

None of these problems was critical, however, and *Columbia* completed its scheduled mission – and more. Landing was initially planned for Edwards AFB as previously, but unseasonably wet conditions there, prompted a diversion to White Sands Test Facility in New Mexico, where in the late 1940s the American space program had taken its first tentative steps. Gusty winds at White

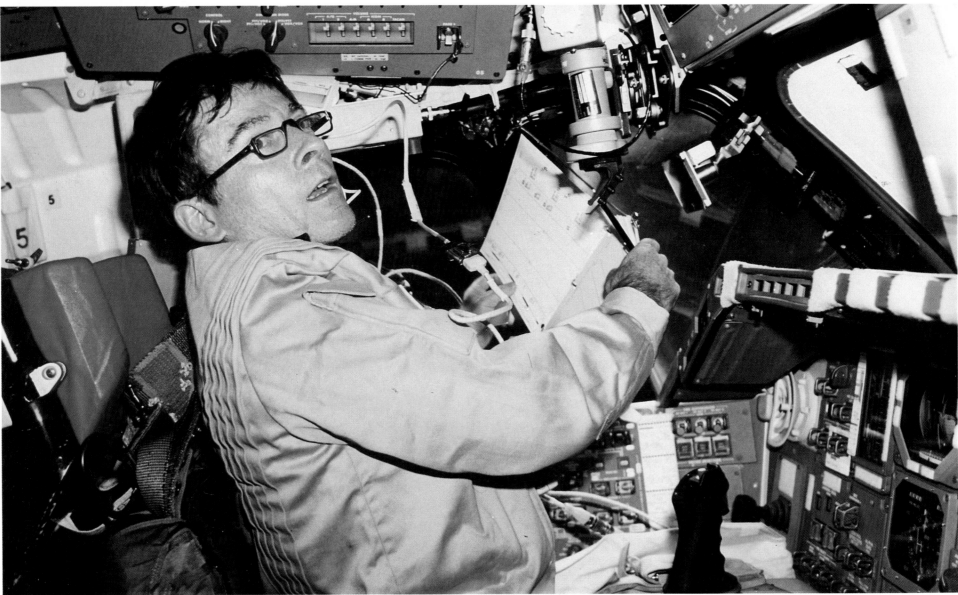

Far left: The solid rocket boosters arc up and over and plummet back to Earth. They parachute back to a gentle splashdown at sea. Here, the recovery ship Liberty returns to Kennedy with one of the boosters.

Bottom left: Columbia has shed its external tank and climbed into orbit. Everything is going like clockwork. Commander Young is well satisfied.

Right: Young is not so happy when he looks through the aft flight deck window toward the engine pod. Some of the tiles are missing. On the pod the absence of a few tiles is not important. But it will be a different matter if tiles are missing underneath.

Below: Fears about the missing tiles are groundless and Columbia glides easily down to make a textbook landing at Edwards Air Force Base in California, on 14 April 1981. The 54 hour maiden flight of Columbia is a triumphant success.

Right: On the evening of 3 November Columbia sits on the pad just hours away from the scheduled lift-off. The clouds raise doubts about weather prospects for the following day. The weather on the morrow is fine, however. But the countdown clock stops just 31 seconds before lift-off and the launch is postponed.

Above: After Columbia has landed, ground crews move in to remove residual fuel and gases that could pose a safety hazard during post-landing operations.

Right: In August 1981 Columbia is hoisted onto the external tank inside the Vehicle Assembly Building as preparations speed ahead for its second journey into space (STS-2), scheduled for November. It has been thoroughly checked and serviced and has had lost and damaged heat-shield tiles replaced.

Sands, however, prevented a landing there until a day later than scheduled. The crew put the extra day in orbit to good use, acquiring extra data from its experimental payload.

The fourth and final shuttle test flight on 27 June 1982 began with a textbook lift-off from Kennedy, followed by a flawless seven-day flight. Appropriately, *Columbia* returned home on 4 July and was welcomed by President Ronald Reagan. The President likened the final, routine test flight of *Columbia* to the 'golden spike' that signaled the beginning of transcontinental railroading in an earlier era. 'Now we can move forward,' he said, 'to capitalize on the tremendous potential offered by the ultimate frontier of space.'

Going operational

On 11 November 1982 the world's first re-usable space transportation system, developed at a cost of some $10 billion, at last opened for business when *Columbia* soared into orbit for the fifth time, exactly on schedule. It was carrying the first commercial cargo, not one, but two communications satellites. It was also carrying the first American four-man crew. Two of them were mission specialists, the first of a new breed of astronauts concerned, not with flying the spacecraft, but with launching the satellites it carried and with general 'housekeeping' duties in orbit.

The launch of each satellite was carried out by mission specialists Joseph Allen and William Lenoir from the payload operations console at the rear of *Columbia*'s flight deck. The satellite was spun out of a protective pod in the orbiter's cavernous 60 foot (18 meter) long payload bay. Next, pilot-astronauts Vance Brand and

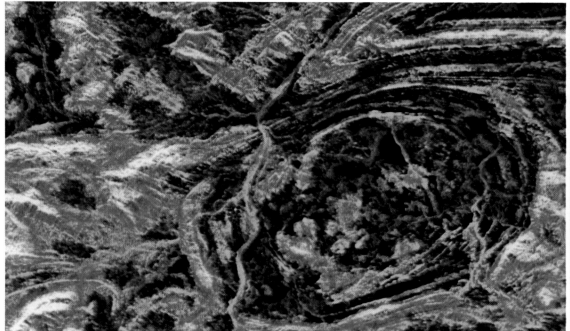

Above: This fascinating false-color image resulted from data acquired by a special radar system flown on STS-2, known as the shuttle imaging radar. It was developed by the Jet Propulsion Laboratory at Pasadena, California. The picture shows part of the Hammersley Mountains in Western Australia. Differences in color reflect differences in topography. Blue represents the smoothest areas, red the roughest.

Right: Astronaut communicator, or CapCom, for the STS-2 flight is Sally Ride, an expert in working the robot arm and scheduled to become the first American woman to be launched into space.

Robert Overmyer maneuvered the orbiter some 20 miles (30 km) away. Then a rocket beneath the satellite ignited and thrust it into a trajectory that would take it into a geostationary orbit some 22,300 miles (35,900 km) high. All satellites destined for high orbits must use a similar technique, since the operational ceiling of the orbiter is only about 500 miles (800 km).

Another commercial payload carried on STS-5 was more modest. It was a 'getaway special', a payload in NASA's small self-contained payload program. Purchased by the Federal Republic of Germany's Research and Technology Ministry, it tested a self-operating oven and an X-ray unit for metallurgical research. NASA offers these special facilities as a means of allowing industries, universities, and even individuals to participate in space research at minimal cost ($3000-$10,000).

Both scheduled satellite launchings went ahead as planned, and the crew, pleased with their work advertised their services on a subsequent telecast with a card reading: 'Satellite Deployment by Ace Moving Co. We deliver.' Indeed NASA had delivered, and

Above: Trails of vapor form at the wingtips of Columbia as it glides down towards the Edwards Air Force Base for its second landing there, on 14 November 1981. Note the dark underside, covered with thick black-coated tiles which bear the brunt of the re-entry heating.

Left: Columbia and the space shuttle go commercial on STS-5, as the communications satellite Anik C-3 spins out of its protective pod in the payload bay in November 1982. Later, the attached PAM (payload assist module) booster will fire to deliver the satellite into geostationary orbit 22,300 miles (35,900 km) above the Earth's equator.

Right: STS-5 mission specialist Joseph Allen working on Columbia's aft flight deck. He is looking up through one of the windows in the top of the fuselage. Behind him is one of the windows that look into the payload bay. The controls in this part of the craft relate to payload activity, such as satellite launching. Note the camera floating in front of him.

These pages: Looking a little worse for wear after completing its fifth journey into space in 19 months, Columbia is pictured in the early morning at Edwards Air Force Base on 16 November 1982. Later it will be flown back to Kennedy to be modified for its next mission in a year's time.

Right: In gleaming pristine condition Challenger, *the second operational orbiter in the shuttle fleet, is towed into the Vehicle Assembly Building to be mated with external tank and boosters. Lift-off for its maiden flight, the shuttle's sixth (STS-6), is scheduled for early 1983.*

Far right: A spectacular nightime view of the shuttle launch pad in February 1983 as work proceeds on readying Challenger *for its space debut. Contained in the payload canister in the rotating service structure (left) is the mission payload, the first tracking and data relay satellite (TDRS).*

earned for themselves some $18 million, their first paycheck from satisfied customers.

NASA itself, however, was far from satisfied with another aspect of the mission, which was to test the new shuttle spacesuit for the first time. Allen and Lenoir were to attempt the first American spacewalk since Skylab. The first attempt had to be postponed because Lenoir went down with a severe attack of space sickness, possibly aggravated by that astronaut's consumption of his favorite spicy hot jalapeño peppers during the flight! When Allen and the recovered Lenoir did eventually don their suits, they malfunctioned. So the planned spacewalk was canceled.

On 16 November *Columbia* returned to Earth, landing as usual at Edwards AFB. It was looking a little the worse for wear. There were pits in its heat-shield tiles and dark streaks across the fuselage, as well as which it had a flat tire. But it had after all traveled around the Earth 384 times and covered a distance of more than 10 million miles (16 million km).

Columbia deserved, and was to be given, a rest. It returned to 'drydock' at Kennedy, where work started immediately on a major refit, to expand its interior to take a larger crew and to up-grade its

engines. When *Columbia* next flew, it would be on another historic mission, carrying into orbit a European space laboratory.

A worthy *Challenger*

Only a few days after *Columbia* arrived at Kennedy to begin its refit, a new orbiter was trundled out to the launch pad, piggy-back on the shuttle stack, looking to a January launch debut. Named *Challenger*, it was somewhat lighter than its predecessor thanks to simplification in design of certain structural elements. It also had modified engines that increased full-throttle power by nearly 10 per cent, compared with *Columbia*.

This engine up-rating caused problems, however. During flight readiness firings on the pad, fuel leaks were detected in all of the engines through hairline cracks in the inner plumbing. The replacement of one engine and repair of the other two pushed back the launch date. An additional delay was caused when a lashing rainstorm at the pad forced fine salty grit into the payload bay and contaminated *Challenger*'s special cargo, the first tracking and data relay satellite (TDRS).

It was not until 4 April 1983, therefore, that NASA's new shuttle

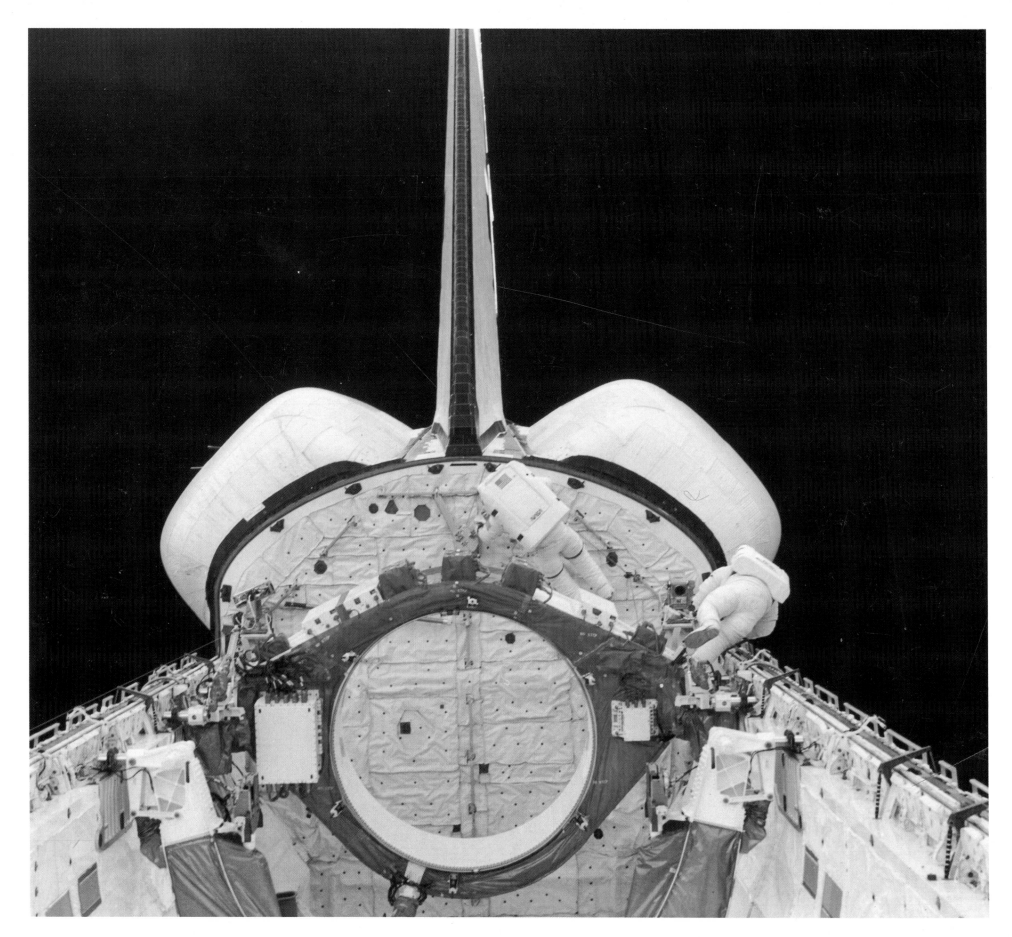

Left: Dwarfed by Challenger's huge tail and engine pods, STS-6 astronauts Story Musgrave (left) and Donald Petersen test the new shuttle spacesuits for the first time in April 1983. A portable life-support system backpack is built into the upper torso of the suits, which therefore no longer need umbilicals.

Right: Story Musgrave provides us with an excellent view of the shuttle spacesuit as he floats alongside the payload bay. For safety he is tethered to a slide-wire system that runs along the edge of the bay.

Above: Sally Ride practices in the payload operations simulator at the Johnson Space Center, Houston, in preparation for her forthcoming flight on STS-7. With Ride is Frederick Hauck, who will be pilot on the mission.

Left: On its second mission (STS-7) Challenger will carry four payloads. They are pictured here loaded in a canister before being taken to the launch pad for installation vertically in the orbiter's payload bay. In this pod at the bottom are two communications satellites. The other two payloads are experiments, the top one (SPAS-1) provided by West Germany.

Right: A beautiful picture of Challenger in orbit, taken from the payload SPAS-1, temporarily launched into space by the orbiter's robot arm. It will later be recaptured and returned to Earth. This, the first deployment and retrieval of a spacecraft, provides further proof of the versatility of the manipulator arm.

orbiter sped into the skies on mission STS-6. It was a perfect launch. The deployment of the TDRS later in the mission was also perfect. This $135 million package of electronic wizardry was to form part of a new communications network that would eventually replace most of the ground tracking stations of NASA's spacecraft tracking and data network (STDN). When fully operational, there would be two TDRS satellites, capable of handling communications from 25 spacecraft, including the shuttle, simultaneously.

A few hours after the TDRS has been deployed on STS-6, the booster rocket beneath it, called the inertial upper stage (IUS), fired to accelerate it towards a geostationary orbit 22,300 miles (35,900 km) high. But the second stage of the IUS cut off prematurely, and the TDRS found itself, uselessly, in a much lower orbit. Signals sent to the TDRS separated it from the IUS and unfurled its two huge solar panels and dish antennae. Telemetry confirmed that the TDRS worked perfectly. The problem was how to get its correct geostationary orbit.

The TDRS problem was one for the ground controllers, not for the STS-6 astronauts. They had other work to do. Most important,

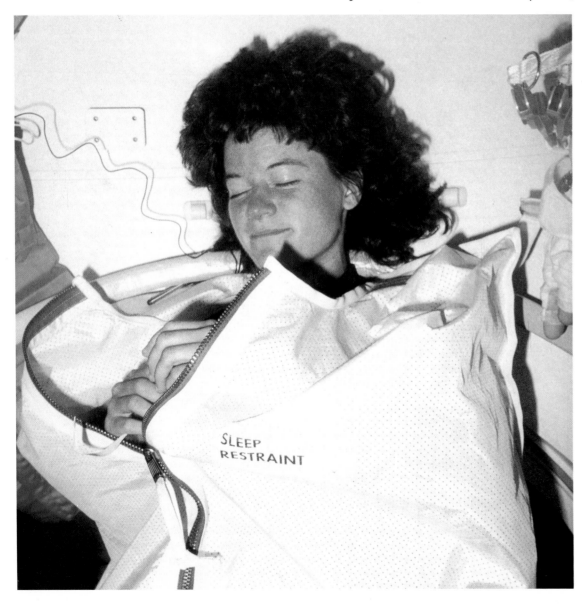

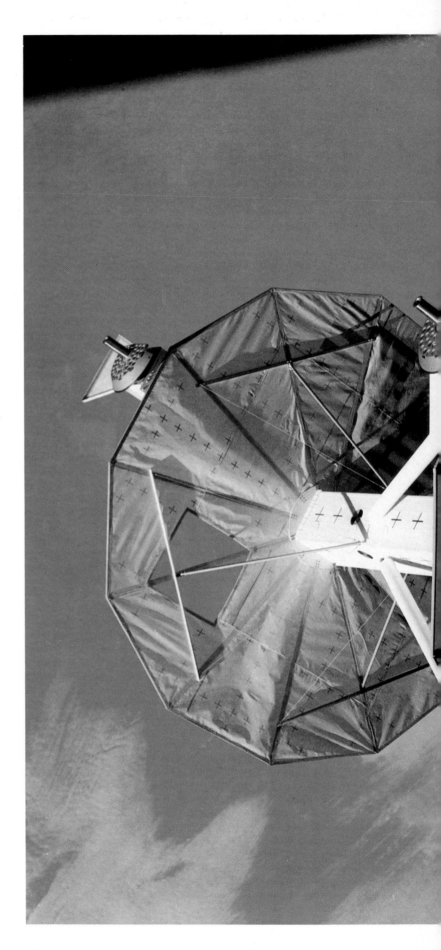

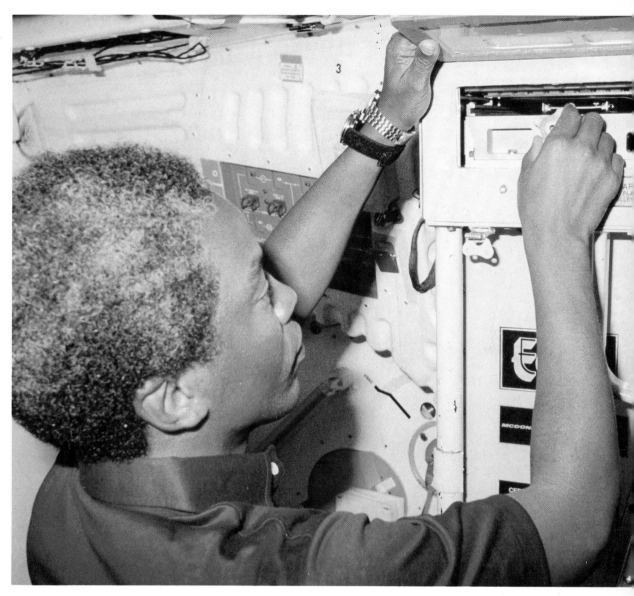

Above: On Challenger's third mission (STS-8) in August 1983, mission specialist Guion Bluford checks the operation of an electrophoresis experiment in the mid-deck area. The experiment seeks to improve the separation of biological materials by means of electric fields.

Left: Further testing of the manipulator arm takes place on STS-8 using this 'cotton reel' as a target. Note the grappling pins located at various points. These are gripped by a so-called 'end effector' at the end of the arm.

the two mission specialists, Story Musgrave and Donald Peterson, were preparing to do what their early colleagues had attempted: that is to test the new shuttle spacesuits.

The crew of a shuttle normally work on board in 'short-sleeve' order. They do not need spacesuits because the orbiter cabin is fully pressurized, like a terrestrial airliner. Only when they leave the cabin to work outside do they need to don spacesuits, technically known as extravehicular mobility units, or EMUs.

The shuttle suit is a great improvement on the Apollo suit that preceded it. Like the earlier model, it is a multilayer garment, but is more comfortable to wear and gives greater mobility in zero-g. It is made up of two parts, an upper torso and trousers. The torso is a rigid structure with a portable life-support system built in. It connects with the trousers by means of a metal ring at the waist. The suit is completely self-contained unlike earlier spacesuits, which were supplied with oxygen, power and so on through an umbilical tube from the spacecraft's on-board life-support system.

On 7 April Musgrave and Peterson suited up in the airlock on *Challenger*'s mid-deck and emerged into the spacious payload bay.

Pausing only to clip their safety tethers on to a wire running along the sides of the bay, they made the first American spacewalk for over nine years. *Time* magazine described this first shuttle EVA graphically. The two astronauts it said 'looked like acrobatic snowmen as they joyously floated, tumbled and wheeled about in their puffy white spacesuits.'

Challenger completed a perfect five-day mission on 9 April with a perfect touchdown at Edwards AFB. In six weeks, after a record turn-around, it was back on the launch pad. On 18 June it was blasting into space again. This flight (STS-7) set two new records. It carried the world's first five-person crew. Five 'person' because the crew included the first American woman astronaut, Sally Ride, as a mission specialist.

Sally Ride was not the first woman to fly in space, but the third. The first two had been Soviet cosmonauts: Valentina Tereshkova in 1963 and Svetlana Savitskaya in 1982.

During the flight Ride had special responsiblity, with fellow mission specialist John Fabian, of operating the remote manipulator arm in a deployment and retrieval test of a satellite, SPAS-1. They operated the arm from the payload-bay console at the rear of the flight deck. They could see the satellite stowed in the bay through the rear windows of the cabin and also on closed-circuit TV. They plucked the satellite out of the bay and placed it in orbit.

Using the satellite as a target, mission commander Robert Crippen, making his second shuttle flight, then practiced approach and rendezvous maneuvers. When these were completed, it was plucked from space by Ride and Fabian and re-stowed in the payload bay. This successful exercise would serve as a rehearsal for a mission planned for the following year: to rescue and repair a defunct satellite known as Solar Max.

The only disappointment on STS-7 was that it landed at Edwards AFB instead of back at the launch site, Kennedy. It had to be diverted there because of low-visibility over Cape Canaveral. For the crew, the disappointment was only fleeting. It certainly did nothing to dampen Sally Ride's enthusiasm for space flight. As she commented: 'It's the most fun I'll ever have in my life.'

Three in a row
After another speedy turnaround, *Challenger* was heading back into orbit again on 30 August, on mission STS-8. The take-off this time was truly spectacular, taking place at night. Pilot Daniel Brandenstein described the astronaut's view of lift-off as: 'like the inside of a bonfire'. The night launch was dictated by the orbital requirements of the Indian satellite that was to be launched. In orbit that satellite was successfully deployed, and within five days was functioning well in geostationary orbit.

During the six-day flight the crew began the first shuttle-to-ground communications tests using the TDRS. This was originally, you will remember, delivered into the wrong orbit on STS-6. But by firing the satellite's tiny attitude control thrusters over a 58-day period, ground controllers had managed to maneuver it into the

Above: On STS-8 Challenger *both lifted-off and touched down at night. This dramatic picture shows the touchdown at Edwards shortly after 12.40 am on 5 September 1983.*

Right: The European-built Spacelab being prepared for its maiden voyage on Columbia *in 1983. Supervising the installation of the equipment in the rear of the picture are West Germany's Ulf Merbold and US astronaut Owen Garriott, a veteran of Skylab. Merbold will become the first foreign national to fly aboard an American spacecraft.*

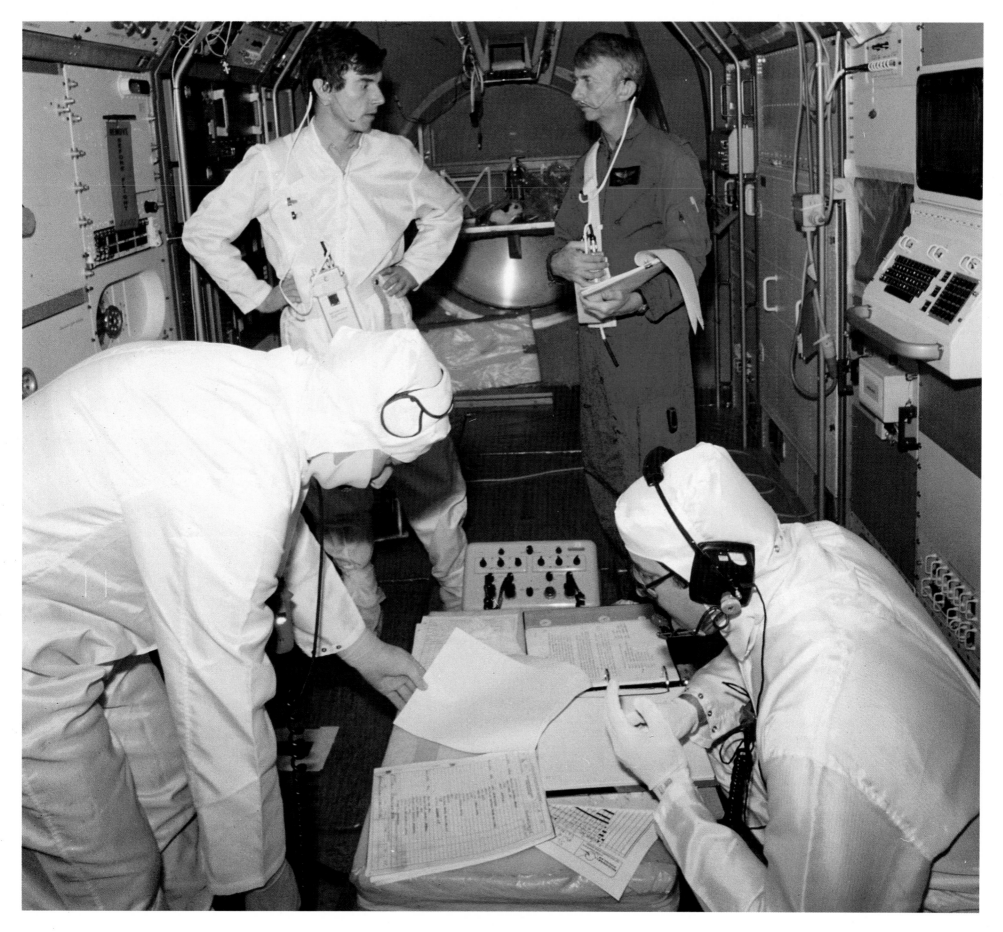

These pages: Looking much refreshed after its well-earned rest, Columbia returns to the launch pad in September 1983 for a late fall launch. Spacelab is already installed in the payload bay. Note the unpainted orange external tank, standard on shuttle missions since STS-3. It is the color of the spray-on insulation applied to prevent the supercold liquid hydrogen and liquid oxygen inside from boiling away.

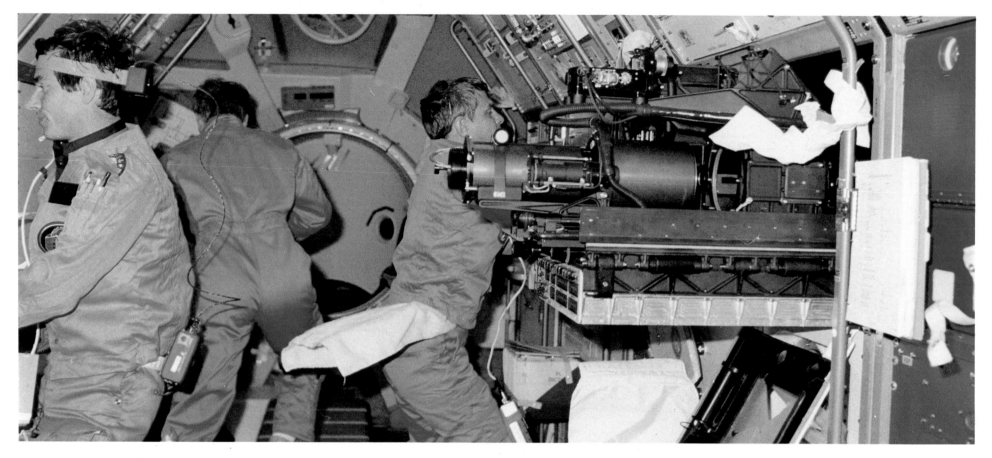

correct orbit above the equator over Brazil. Tests early in the mission were successful, but later ones failed. This was worrying because the success of the next flight would depend on the TDRS system working.

On STS-8 *Challenger* not only lifted-off at night, it also landed at night, as before at Edwards AFB. To spectators it was an eerie experience as the orbiter suddenly and silently emerged from the night and swooped steeply down to the brightly-lit runway.

Europe's orbiting laboratory
The ninth mission of the space shuttle was going to be something special. It was to carry a European-built space laboratory into orbit. Named Spacelab, it had been built in Germany and financed to the tune of $1 billion by ten members of the European Space Agency (ESA).

The STS-9 mission would also break new ground because of the presence on board of two non-astronauts, one of whom was European. He was Ulf Merbold, a West German physicist. Merbold and the other non-astronaut, American biomedical engineer Byron Lichtenberg, were to be payload specialists. Their duties on board Spacelab would be to tend the scientific equipment and carry out experiments. 'Old timers' among the other members of the record six-man crew, would be John Young, attempting a world-beating sixth flight in space; and Owen Garriott, a veteran of Skylab.

Spacelab was the product of a collaboration between NASA and the European Space Research Organization (ESRO, the forerunner of ESA) that dated back to 1972. Since 1973 it had been an integral part of the space shuttle program, and, like the shuttle itself was

designed to be re-usable. The first parts of Spacelab began arriving at Kennedy in late 1981. It was installed in orbiter *Columbia* in August 1983. Preparations were then well underway for an end-October launch.

However, a routine checkout of the shuttle on the launch pad revealed a defective nozzle on one of the solid rocket boosters. It had to be replaced, which meant rolling the whole stack back to the Vehicle Assembly Building, dis-assembly, replacing the nozzle, re-stacking, and rolling out again. It was the first time that the boosters had presented problems, and it was not to be the last. On the next occasion, however, only a little over two years later, there would be fatal consequences.

The booster problem with *Columbia* put back the Spacelab launch until 28 November and also put some of the experiments in jeopardy. Within a few hours of an uneventful launch, the scientist-astronauts in the crew opened the hatch into a tunnel that led to the space laboratory in the open payload bay. Spacelab is a versatile hardware system and can be flown in a number of configurations. On STS-9 it comprised a long module and pallet. The long module consisted of two identical segments joined together, and measured about 23 feet (7 meters) long and 13 feet (4 meters) in diameter. It formed the pressurized laboratory in which the astronauts worked. The pallet was a U-shaped platform carrying a variety of instruments.

Guinea pigs in space
There were on board the ingenious Spacelab 1 a total of 72 experiments, grouped into five main fields – space medicine,

Above: Up in orbit in November 1983 Spacelab astronauts work round the clock to complete their heavy workload, which includes over 70 different experiments. Engrossed in their work here are, from the left, Ulf Merbold, Byron Lichtenberg and Robert Parker. Merbold wears a headband device to acquire data about the nature of space adaptation syndrome, colloquially called space sickness.

Right: The Spacelab shuttle flight (STS-9) carries a record six-man crew, posing here together for a topsy-turvy portrait. Clockwise from the top of the picture the astronauts are Byron Lichtenberg, Robert Parker, John Young (making his sixth space flight), Ulf Merbold, Owen Garriott and Brewster Shaw.

astronomy, space physics, Earth physics and materials science. The experiments were manned round the clock by the crew, working 12-hour shifts in turn. Instructing and advising the crew from the ground were 200 scientists scattered around the world. Communications link-ups between Earth and Spacelab were complex and suffered breakdown on more than one occasion, resulting in the loss of considerable amounts of experimental data. The major culprit was the accident-prone TDRS, which couldn't live up to its design specifications.

In space medicine the Spacelab scientists carried out intensive investigations into the effects of weightlessness on the human body. This leads in approximately 50 per cent of astronauts to space sickness, more properly termed space adaptation syndrome. The investigations included regular blood sampling, electric shock treatment, and rides in a sled-like contraption. Ironically, this crew didn't suffer space sickness, though they began to feel queasy as a result of the experiments!

Communications problems aside, Spacelab 1 performed exceedingly well, and with distinction achieved the mission objectives of exercising the hardware, flight and ground systems, and the ability of the flight crew to conduct advanced scientific research in space. So delighted were NASA with Spacelab's performance that they decided to extend its 9-day mission by 24 hours.

On the ninth-day in orbit, however, their delight turned to alarm. *Columbia* fired its thrusters as part of the preliminary maneuvers for descent and re-entry, and the burn jolted the craft, making a noise, Young said, like a 'howitzer blast going off in your backyard.' Immediately, one of the orbiter's computers shut down, but fortunately another took over straightaway. A few minutes later came another burn, another jolt and the shutdown of the second computer. Two of the orbiter's four main computers were now unservicable.

The situation was becoming alarming because any further computer failures during the other maneuvering and retrobraking burns could put the crew in mortal danger. Without computers the orbiter cannot fly because it is far too complex a machine. Even when the astronauts fly it manually via the joysticks, all instructions have to be routed through the computers. Fortunately, none of the remaining computers failed, and Young made a textbook landing at Edwards AFB. The NASA philosophy of multiple redundancy in the computer system had paid of. *Columbia*, however, suffered a final indignity when it was found that it had been slightly on fire on landing!

The launch and successful operation of Spacelab — electronic failures and the odd fire aside — marked a significant milestone in shuttle development, and emphasized the outstanding benefits of shuttle operations. The orbiter proved that it can transport into orbit large, complex and sensitive payloads, beyond the capability of conventional expendable rockets. It can carry a large crew, some of whom need not be test-pilot type astronauts of the old school. Because the shuttle launch is carefully controlled, acceleration-wise, astronauts experience only mild discomfort under a maximum of 3gs. As long as they are reasonably fit, ordinary people can venture into space after only a few months' acclimatization training.

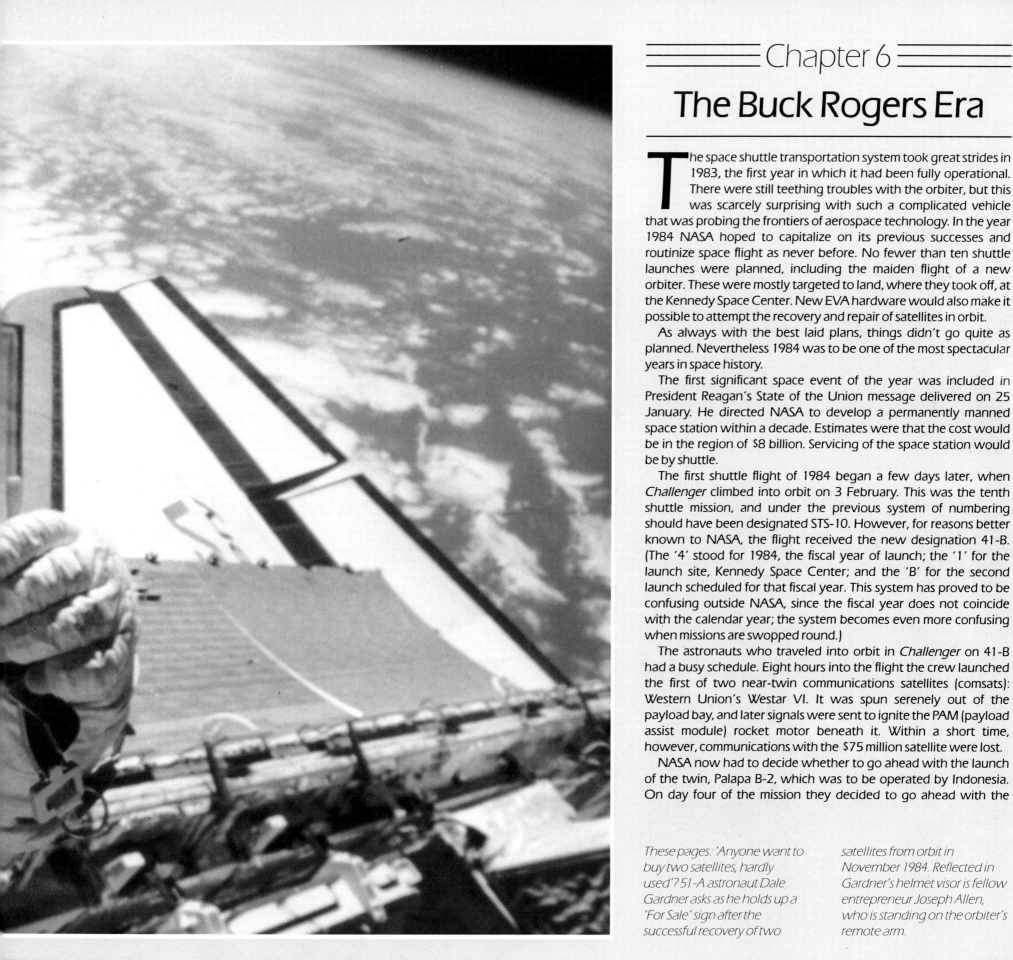

Chapter 6

The Buck Rogers Era

The space shuttle transportation system took great strides in 1983, the first year in which it had been fully operational. There were still teething troubles with the orbiter, but this was scarcely surprising with such a complicated vehicle that was probing the frontiers of aerospace technology. In the year 1984 NASA hoped to capitalize on its previous successes and routinize space flight as never before. No fewer than ten shuttle launches were planned, including the maiden flight of a new orbiter. These were mostly targeted to land, where they took off, at the Kennedy Space Center. New EVA hardware would also make it possible to attempt the recovery and repair of satellites in orbit.

As always with the best laid plans, things didn't go quite as planned. Nevertheless 1984 was to be one of the most spectacular years in space history.

The first significant space event of the year was included in President Reagan's State of the Union message delivered on 25 January. He directed NASA to develop a permanently manned space station within a decade. Estimates were that the cost would be in the region of $8 billion. Servicing of the space station would be by shuttle.

The first shuttle flight of 1984 began a few days later, when *Challenger* climbed into orbit on 3 February. This was the tenth shuttle mission, and under the previous system of numbering should have been designated STS-10. However, for reasons better known to NASA, the flight received the new designation 41-B. (The '4' stood for 1984, the fiscal year of launch; the '1' for the launch site, Kennedy Space Center; and the 'B' for the second launch scheduled for that fiscal year. This system has proved to be confusing outside NASA, since the fiscal year does not coincide with the calendar year; the system becomes even more confusing when missions are swopped round.)

The astronauts who traveled into orbit in *Challenger* on 41-B had a busy schedule. Eight hours into the flight the crew launched the first of two near-twin communications satellites (comsats): Western Union's Westar VI. It was spun serenely out of the payload bay, and later signals were sent to ignite the PAM (payload assist module) rocket motor beneath it. Within a short time, however, communications with the $75 million satellite were lost.

NASA now had to decide whether to go ahead with the launch of the twin, Palapa B-2, which was to be operated by Indonesia. On day four of the mission they decided to go ahead with the

These pages: 'Anyone want to buy two satellites, hardly used'? 51-A astronaut Dale Gardner asks as he holds up a 'For Sale' sign after the successful recovery of two satellites from orbit in November 1984. Reflected in Gardner's helmet visor is fellow entrepreneur Joseph Allen, who is standing on the orbiter's remote arm.

137

launch, expecting and hoping that bad luck wouldn't strike twice. Unfortunately it did.

It was not, however, the satellite electronics that were at fault, but the PAM rockets, which failed to impart sufficient thrust to boost the comsats into the desired geostationary orbit, 22,300 miles (35,900 km) high. Tracking stations later pinpointed the two satellites in elliptical orbits that took them only out to a maximum of some 750 miles (1200 km).

The 'flying armchair'

While NASA on the ground began inquiring into the probable causes of the loss of $150 million worth of satellites, the crew of 41-B had other things on their minds. In particular they were preparing to test-fly one of the remaining items of hardware in the space transportation system: the manned maneuvering unit (MMU). Earlier versions of this one-man propulsion unit had been test-flown in the cavernous forward compartment of Skylab, but never in space.

The MMU has been described as a 'flying armchair', which it vaguely resembles. Its function is to provide transport for an astronaut on EVA. The astronaut backs into the MMU in his spacesuit and straps himself in. Then, by firing thruster jets on the MMU, he can travel around in any direction he chooses. Hand controls on the MMU arms fire the jets. Twin tanks in the back of the MMU provide a supply of compressed nitrogen gas to the thrusters which is sufficient for about 6 hours' operation. Batteries power the controls and flashing locator lights.

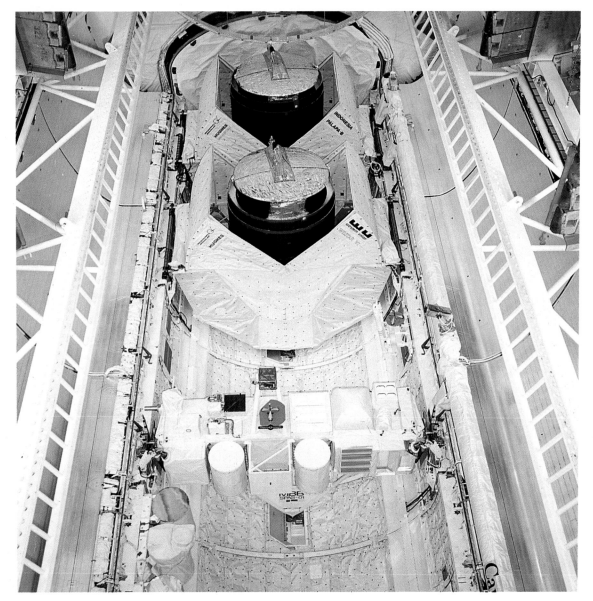

On 7 February 1984 41-B mission specialist Bruce McCandless emerged from the shuttle airlock and donned one of the two $15 million jet-propelled backpacks stored just inside the payload bay. Gingerly he fired the MMU's thrusters, which puffed out gas in one direction and pushed him gently in the other. Soon he was heading out of *Challenger's* payload bay into free space. Completely untethered, he became the first human satellite, jetting away from the safety of his spacecraft to a maximum distance of some 300 feet (96 meters). What if then his thrusters had stuck on? What if his suit had sprung a leak? The dangers were obvious and really didn't bear thinking about. But everything went perfectly. McCandless slowly drifted back and, maneuvering in front of the orbiter's windshield, asked mission commander Vance Brand: 'Hey, you going to want the windows washed while I'm out here?'

After his flight, McCandless turned the MMU over to Robert Stewart who was on EVA with him. On a second spacewalk the following day McCandless and Stewart practiced procedures that were going to be employed on the next flight in an attempt to

Above: A dampener to the 41-B mission, however, is the loss of two very expensive communications satellites, Westar VI and Palapa B-2. They are seen here during installation in Challenger's payload bay prior to the flight. This loss will later trigger a daring, and ultimately successful recovery mission.

Right: Robert Stewart takes a turn with the MMU during the 41-B mission. The MMU proves a great success, and yet another component in the shuttle's hardware is ready.

retrieve and repair a satellite called Solar Max. They also checked out a platform that fixed on to the end of the remote manipulator arm. Called a mobile foot restraint, or more colloquially a 'cherry picker', it was designed to provide a stable platform for astronauts to work on and one that could carry them around the payload bay.

Return to base

The demonstrations of the MMU were spectacularly successful and augered well for the planned expansion of EVAs in the coming months. McCandless and Stewart obviously reveled in their spectacular untethered spacewalks. Commented mission commander Vance Brand: 'They call each other Flash Gordon and Buck Rogers,' referring to two space heroes of the fictional kind.

On 11 February the real-life space heroes headed for home. This time the weather at the Kennedy Space Center was favorable to attempt the first landing there. *Challenger* made its de-orbit burn on orbit 127 over the Indian Ocean off the west coast of Australia. Its path took it across the Pacific, making landfall over the Americas at the Baja Peninsula. Slowing and descending all the time, it

crossed Mexico and Southern Texas and entered Florida north of Sarasota traveling almost due east.

All across central Florida *Challenger* trailed a double sonic boom, that became louder as it approached Kennedy. It went subsonic as it passed over the Center and circled out to sea before turning back and making its final approach on to the 3 mile (5 km) long runway. Guided in by a sophisticated microwave landing system, it made a perfect touchdown at 7.17 am EST. It was two minutes early.

Challenger's triumphant return to Kennedy was the first of any spacecraft to its launch site. It saved precious days of turn-around time and a $1 million bill for transportation from an alternative landing place such as Edwards AFB, site of all but one of previous shuttle landings.

Grappling with Solar Max

The next shuttle mission, 41-C, saw *Challenger* ascend on 6 April directly into the highest shuttle orbit yet, at 290 miles (460 km). Its orbital maneuvering system (OMS) engines were used only once to circularize the orbit, conserving fuel for the maneuvers that would

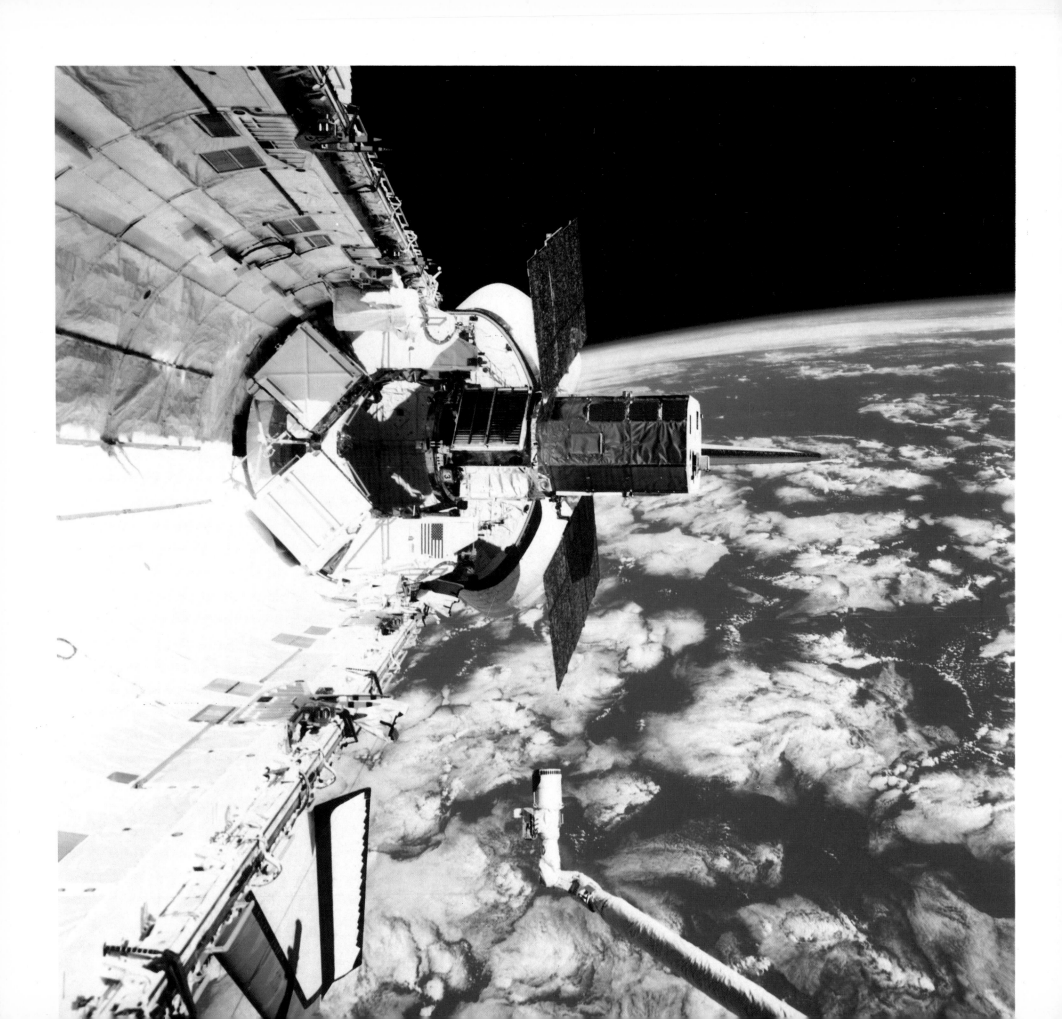

be needed for the main mission objective – the recovery and repair of the satellite Solar Max. Recovery and repair of the satellite were possible because it was built with maintenance in mind. It had a modular design, so that individual parts could be replaced if they malfunctioned. It also had a grapple fixture which could be grabbed by the orbiter's remote manipulator system's robot arm for capture.

Solar Max (short for solar maximum mission) was a satellite that had been launched four years previously to monitor activity on the Sun during a peak period for solar flares and sunspots. After 10 months' operation, however, Solar Max blew three fuses in its attitude control module. This meant that its instruments could no longer be pointed accurately. Later an electronics box serving some of the instruments also failed. The satellite was nearly, but not quite, dead. Using a back-up attitude control system of on-board magnets, engineers at Goddard Space Flight Center's communications and tracking control center were able to keep the satellite's solar panels pointing at the Sun, while it spun round once every six minutes.

The plan of action on the mission called for *Challenger* to rendezvous with Solar Max and for one of the astronauts to fly to it and stop it spinning. Then *Challenger* would move in closer and pluck the satellite from orbit and stow it in the payload bay.

On the third day of the mission, therefore, *Challenger* 'parked' about 200 feet (60 meters) away from Solar Max, and George Nelson jetted across to it in an MMU. Fixed to the arm of the MMU was a device that was designed to latch on to a pin protruding from the satellite. The idea was that Nelson would, while dodging the flailing solar panels, dock with the satellite and fire the jets of the MMU to stop it spinning.

Nelson, unfortunately, was out of luck. He couldn't dock with the satellite, and his attempts to do so started to make it wobble. Mission commander Robert Crippen's suggestion, that Nelson try to steady the satellite by grabbing one of the solar panels, just made matters worse. Solar Max began tumbling unpredictably around all three axes. After futile attempts to grab the gyrating satellite with the robot arm, *Challenger* backed away.

Over the next two days Goddard managed to steady the

satellite again and *Challenger*, now running low on fuel for maneuvering, moved in to attempt another capture. This time it was successful. The robot arm latched onto the grapple fixture and lowered Solar Max into a cradle in the payload bay. The next day Nelson and fellow astronaut James van Hoften emerged into the bay to carry out running repairs. Despite their bulky spacesuits and gloved hands, they dextrously unscrewed panels and cut and taped up wires. In much less time than expected, they had replaced the faulty attitude control module and electronics box. Next day *Challenger*'s robot arm lifted Solar Max out of the payload bay and replaced it into orbit, working perfectly.

The Solar Max mission program, which had cost over $300 million to bring to fruition, was again back in business.

'We have an abort'

The successful recovery and repair of Solar Max was yet another demonstration of the versatility of the shuttle system. Well did the

41-C astronauts dub themselves the 'Ace Satellite Repair Co.' With the next shuttle launch, 41-D, NASA would have another orbiter named *Discovery*, becoming operational. With three operational orbiters and another to follow in 1985 therefore, they could look forward to an accelerated flight program that would help clear the backlog of satellite launchings occasioned by the technical failures that had fouled up earlier schedules. Or so they thought.

So it was with high hopes that launch control counted down *Discovery* on 25 June. The countdown proceeded uneventfully as far as the T -9 minute hold. Then the back-up computer on board malfunctioned, and the launch had to be postponed. But it was a minor fault, and next day the countdown was resumed. This time it came out of the T -9 minutes hold, and six seconds before lift-off time the main engines fired, on schedule. But two seconds later they sputtered out. 'We have an engine cut-off,' said the NASA commentator, 'We have an abort.' Ominously, unburnt hydrogen gas exploded beneath the engines, but fortunately the blaze was

Below: With seemingly no limit to their versatility, the 41-C astronauts prove brilliant photographers, taking some of the most riveting Earth photographs of the shuttle era. Here they snap the island of Hawaii. Smoke rises from the lava flowing from the erupting volcano of Mauna Loa.

Right: Over Algeria the astronauts photograph this barren desert region virtually inaccessible overland. In the upper right are the sand dunes of the Grand Erg Oriental.

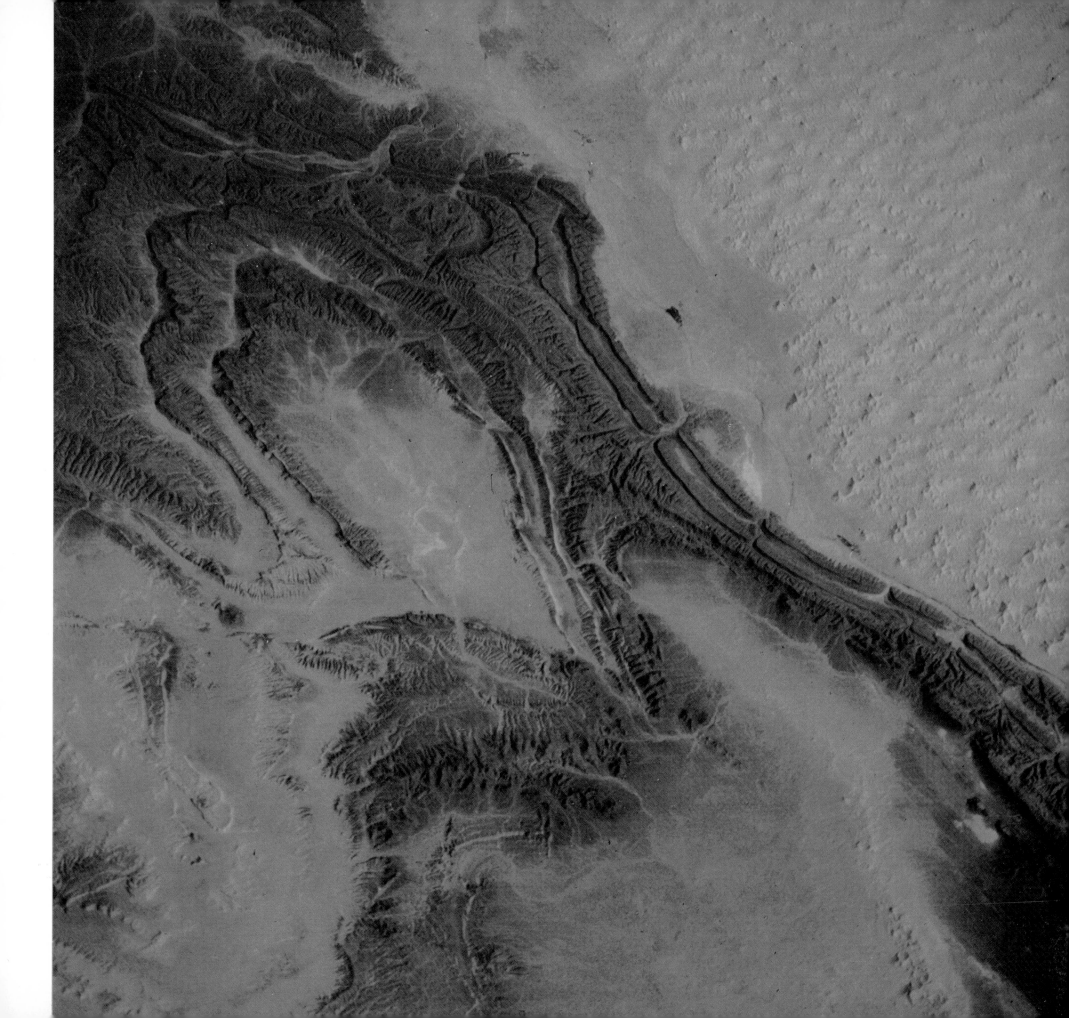

Left: This orbital photograph shows the great fan of the fertile Nile delta in Egypt. Top right is the Mediterranean Sea. To the right of the delta is a waterway that signals its artificial nature by being dead straight, compared with the zig-zagging channel of the Nile itself. The waterway is the famous Suez Canal, which links the Mediterranean with the Red Sea.

Above: The third shuttle orbiter, Discovery, flies into the Kennedy Space Center atop its 747 carrier in November 1983. After separation, Discovery will be towed the 5 miles (8 km) to the orbiter processing facility to be prepared for its maiden flight into space in the summer of 1984.

quickly doused. If it hadn't, who knows what might have happened? The six astronauts were still strapped in their seats sitting on millions of pounds of high explosive fuel.

This fault was far from minor and required two months' work to put right. NASA's schedules were once more in disarray and their credibility again suspect. If they were to succeed in the commercial space launching business, they had to deliver payloads accurately (not lose them, as happened with the Westar and Palapa satellites on 41-B) and what is more, on time. *Discovery* tried to lift off again on 28 August, but was grounded by a flaw in the flight computer software. Next day, however, it at last got airborne. Orbiter number three was operational. Kennedy launch director Bob Sieck sighed with relief. 'The launch team is ecstatic,' he said.

The frisbee satellite

The crew of 41-D, however, had little time for ecstasy as they got to grips with the heaviest workload yet on a shuttle flight. During their six-day mission they launched a record three satellites. Two had the same kind of PAM boosters that had failed in February. This time, however, they fired correctly, boosting the satellites into geostationary orbits. These two satellites were deployed vertically from cradles in the payload bay, as on previous missions. However, the third satellite deployed, Leasat-1, was rolled out sideways from the payload bay like a frisbee. It was the first satellite to be

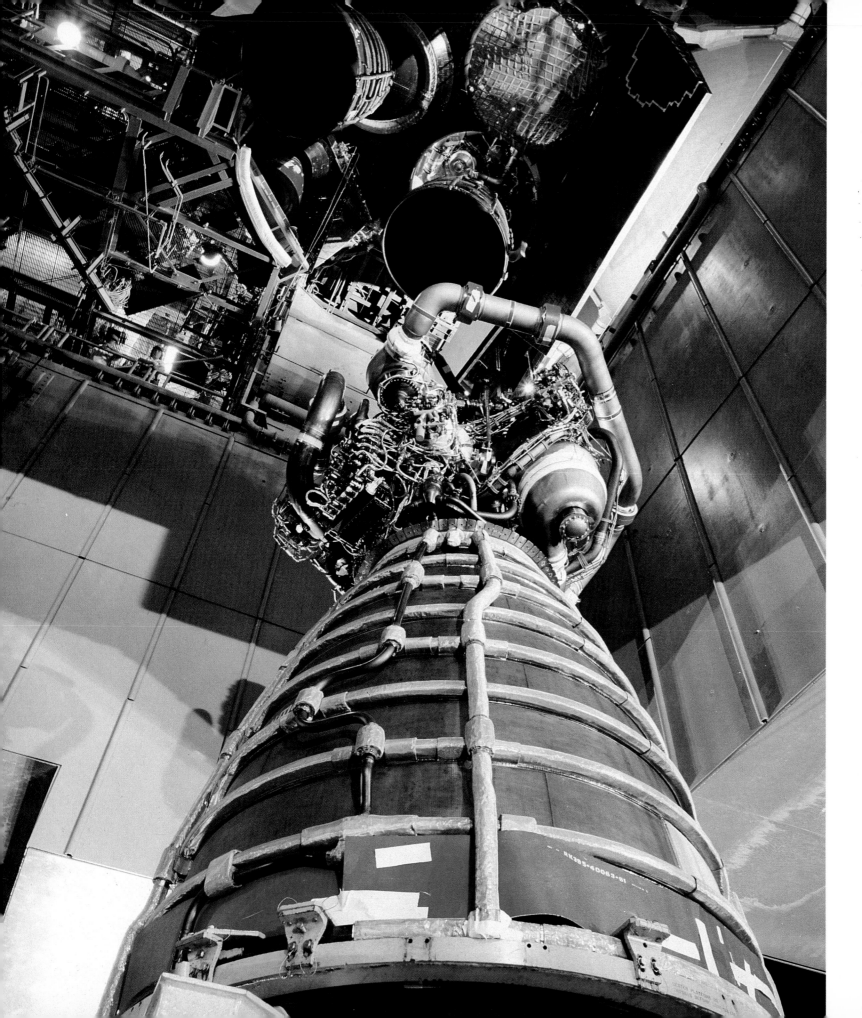

Left: Discovery *suffers a launch abort on 26 June 1984. Just four seconds before lift-off, a fault is detected in main engine number 3. It is rolled back to the Vehicle Assembly Building, where the faulty engine is removed and replaced.*

designed specially for launching from the shuttle. That too soared successfully to its intended high orbit.

On board 41-D was Judy Resnik, America's second woman in space, who was destined tragically never to make it into space again. Her responsibility on the mission was to deploy a dummy foldaway solar panel. This was a prototype solar array that could be used on future space stations. As Resnik activated the controls, the solar panel slowly unfolded, concertina-like, from a housing only a few inches deep and extended to a length of more than 102 feet (31 meters). It was one of the largest structures ever deployed in space.

Another highlight of the flight was the inclusion of Charles Walker as the first commercial payload specialist. McDonnell Douglas picked up a tab of $80,000 for Walker's ride on the shuttle with a drug-processing unit designed to manufacture a 'secret' new hormone. It made use of the zero-g environment to separate and purify chemicals hundreds of times faster than is possible on Earth. The advantage of having a human presence for such work was demonstrated yet again when the drug-processing unit's

Below: Scattering flocks of birds on a nearby lagoon, Discovery accelerates from the launch pad on 30 August 1984, airborne at last on its fourth attempt. It is the first orbiter to make extensive use of lightweight thermal blanket material on non-critical areas of its exterior. It is shuttle mission 41-D.

control computer started playing up. Walker managed to bypass the computer and operate the machine manually.

The feminine touch
The next shuttle flight in October (mission 41-G) was again to be made by *Challenger*, and was the 13th of the progam. It was far from being an unlucky one and indeed set some impressive new records. It had a record crew of seven, two of whom were women. One was Sally Ride, becoming the first American woman to make two space flights. The other was Kathryn Sullivan, who became the first American woman to go spacewalking. She had just been pipped at the post for being the world's first female spacewalker by Russian cosmonaut Svetlana Savitskaya, who had helped carry out repairs of the Russian space station Salyut two months previously.

On their EVA Sullivan and fellow spacewalker David Leestma practiced procedures for in-flight servicing of satellites. Working inside the payload bay, they coupled up a hose between a hydrazine fuel tank and a dummy satellite tank. Later, fuel was transferred between them by remote control. This demonstration

served to emphasize the shuttle's service role in orbit, that of repairing and extending the life of satellites that malfunctioned or ran out of fuel.

Sally Ride's duties on the mission, as before, centered on manipulation of the robot arm. She used it to launch a satellite, giving it a good shake in an attempt to free its frozen solar panels. This didn't work, however, and she had to wait for the Sun's heat to thaw them out before she could release the satellite into space. Later her assistance was requested when the antenna of a radar scanning device failed to deploy. With unerring accuracy, she gave the antenna a good clout. It sprung open. In space, as on Earth, as Skylab 2 astronaut Charles Conrad had discovered earlier, when all else fails, thump it!

Space-age salvage

For the final mission of 1984, 51-A, NASA turned to some unfinished business. In February it had lost two expensive satellites, Westar VI and Palapa B-2. The companies that had insured them had paid out some $180 million compensation to the owners. So, underwritten by the insurers, the astronauts would

attempt a daring salvage mission: chase, rendezvous with and capture the errant satellites, and bring them back to Earth.

Prior to the launch of 51-A, the satellites would be maneuvered into identical orbits accessible to the shuttle. The problem was, how to grip hold of them for recovery. The satellite recovered successfully in April, Solar Max, was designed with a grapple device especially for recapture. The Westar and Palapa satellites on the other hand, had no such fixtures.

Fortunately, one of the astronauts who would be involved in the salvage operation, Dale Gardner, came up with an ingenious idea. He suggested using a device that could be inserted into the propulsion motor of each satellite and then lock. The device would also carry a grapple pin that the robot arm could grasp. From his sketch on the back of an envelope evolved what NASA verbosely called an apogee kick motor capture device. From its shape it soon came to be nicknamed 'the stinger'.

Riding the stinger

After a one-day delay because of high winds, *Discovery* lifted off on mission 51-A on 8 November. Over the first three days the crew

Above: In charge of the solar array experiment on 41-D is Judith Resnik, the second American woman to become a fully fledged astronaut. As you can see, her long tresses float free, Afro-style, in the weightless environment.

Right: The most interesting task on 51-A is the deployment of a huge mock-up solar-cell array. When fully extended, it reaches out to a distance of 105 feet (32 meters). It is one of the biggest structures ever deployed from a spacecraft.

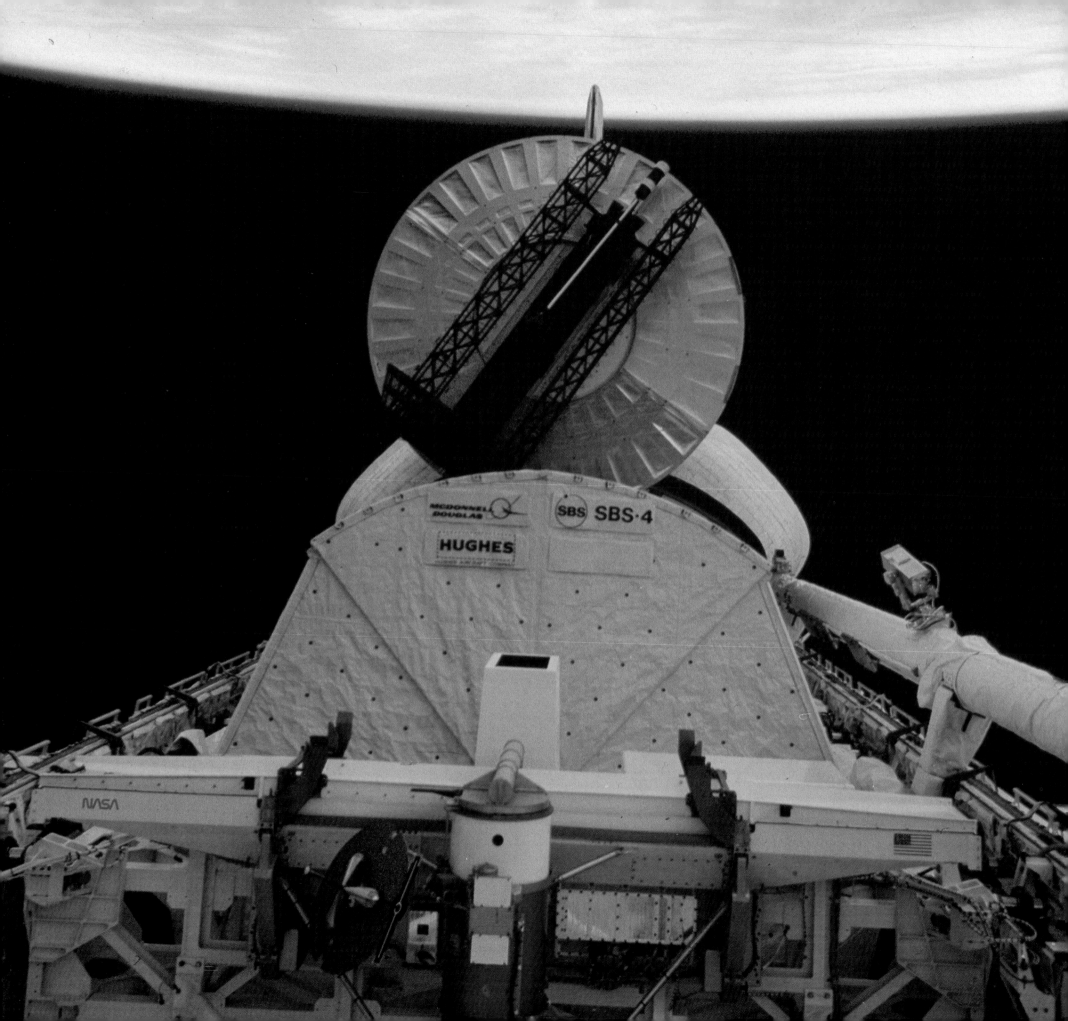

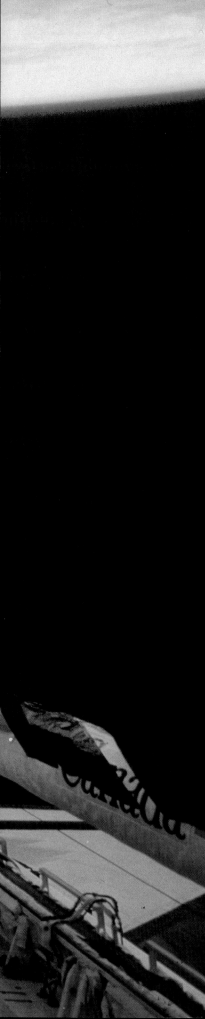

Far left: Discovery launches a record three satellites on its maiden voyage. Here it is deploying a Syncom/Leasat-2 satellite by the 'frisbee' method, rolling it from the payload bay. Interestingly in the foreground, is the container that holds the solar-cell array. It measures only a few inches deep, but look at the size of the array! (see page 153).

Left: It's going to be very crowded in Challenger *when these seven astronauts go aloft on shuttle mission 41-G. They are leaving the operations and checkout building at Kennedy, where the crew quarters are, heading for the launch pad for a 7 am lift-off on 5 October 1984. Heading the procession of astronauts is Kathy Sullivan. Behind her in order come Robert Crippen, Paul Scully-Power and Jon McBride. The right-hand column is led by Sally Ride about to make her second space flight. Behind her are David Leestma and Canadian payload specialist Marc Garneau.*

launched two satellites and started a crystal-growing experiment. Then on day four they chased the first of the runaway satellites, Palapa, and maneuvered to within 35 feet (11 meters) of it. The three main stars in the performance that was to follow were Gardner and fellow spacewalker Joseph Allen on the outside, and mission specialist Anna Fisher on the inside manning the robot arm.

While Gardner anchored himself in footholds on the edge of the payload bay, Allen donned an MMU, picked up the stinger from its stowed position and jetted with it over to the satellite. He inserted it easily into the motor, locked on and then fired the thrusters of the MMU to stop the satellite rotating. Fisher then reached up with the robot arm and, with Allen's assistance, grabbed the stinger's

grapple pin with its claw. She next drew the satellite over to the payload bay. Then Allen tried to fix another grapple fixture to the top end of the satellite, but it didn't fit. This meant that the robot arm couldn't be used for holding the satellite during the rest of the berthing maneuvers. There was only one solution – human muscle power. And so, proving that weightlessness has its uses, Allen held the satellite aloft for a whole Earth orbit (90 minutes), while Gardner removed the stinger from it and attached a berthing device. Then they manhandled the satellite into its berth and secured it.

The following day Allen, Gardner and Fisher rested while *Discovery* chased the second rogue satellite, Westar VI. Then it was

back to work for the second recovery attempt. This time Gardner rode the stinger, docked with Westar and stopped it rotating. Then Allen, riding on the robot arm, grasped the satellite, while Fisher hauled him in. Gardner flew back, removed the stinger, and they both wrestled the satellite into its berth and secured it. 'Houston,' reported mission commander Frederick Hauck, 'we've got two satellites locked in the bay'.

NASA was jubilant at the success of the capture. So were Lloyds of London, who had borne the greater part of the insurance burden when the satellites went astray. The managers there instructed that the famous Lutine bell should be rung twice, a traditional celebration of a successful, but usually terrestrial, salvage. As soon as *Discovery* touched down at Kennedy, the recovered satellites were whisked away for renovation. Later they would be offered for sale, the first secondhand satellites to appear on the market, priced about $35 million each — a real bargain.

Left: David Leestma (left) and Kathryn Sullivan work in the aft section of the payload bay, carrying out the simulated refueling of a satellite propellant tank. Sullivan is making the first ever spacewalk by an American woman.

Right: 41-G commander Robert Crippen is in position on Challenger's port-side flight station during re-entry into the atmosphere on 13 October 1984. Outside is the ruddy glow caused by re-entry heating, which raises the temperature of some parts of the spacecraft to more than 1500°C (2700°F). Re-entry is one of the most dangerous parts of every mission, when the crew are in a communications blackout.

Left: On 8 November 1984 Discovery (on mission 51-A) lifts off for one of the most ambitious missions yet attempted, which includes attempts to recover the two satellites that were lost after launch the previous February, Westar VI and Palapa B-2. Here Dale Gardner, flying an MMU and riding a 'stinger', is jetting over to the gently rotating Westar.

Above: Dale Gardner has hard-docked with Westar and will now fire his MMU jets to stop it spinning. Then he will jet back towards Discovery.

Senator in orbit

The first of the shuttle flights in 1985 (51-C) was a classified military mission for the Department of Defense. During a nominal four-day mission *Discovery* launched a state-of-the-art communications satellite codenamed Aquacade into geostationary orbit. The satellite was a Sigint (signals intelligent) spacecraft designed to eavesdrop on Soviet radio, microwave, telex and satellite communications.

No Buck Rogers' style astronauting was required on that mission but there was on three of the eight other shuttle missions that took place in 1985. The first was unscheduled.

After 51-C preparations went ahead to launch *Challenger* in February on mission 51-E. Then the snarl-ups started. The TDRS (tracking and data relay satellite) scheduled as the payload, developed a fault, and 51-E was canceled. Most of the crew were switched to the next scheduled shuttle, 51-D, and payloads were swopped round. The orbiter would be *Discovery*, and would be launched in late March. But on 8 March, while *Discovery* was being prepared in the orbiter processing facility, a work platform

159

fell onto one of the payload-bay doors and put a hole through it. As a result of the accident, during which a technician broke a leg, the launch date slipped until 12 April, four years to the day since the first shuttle flight.

Discovery's launch on that anniversary day was perfect. Included in the seven-member crew were US Senator Jake Garn, chairman of a Senate subcommittee that holds NASA's purse-strings. He had been critical of the decision to go ahead with a manned space station, favoring first the fuller exploitation of automation and robotics in space. As it turned out he was to be given a salutary lesson on the advantages of a human presence in space. Though undoubtedly present on the shuttle for public relations reasons, Garn nevertheless volunteered to act as a guinea pig in biomedical experiments, helped by mission specialist astronaut Rhea Seddon, who was also making her first space flight.

As part of the experiment program, the Senator was wired up with microphones to record the sound of his digestive processes in action. The physicians monitoring the experiment got more than they bargained for when Garn experienced two days of space sickness!

Flyswatting

The first satellite launch of 51-D went without a hitch on day one of the mission. The Canadian Anik communications satellite spun from the pod in the payload bay, and three-quarters of an hour later was rocketing up to high orbit. On day two a communications satellite leased to the US Navy called Leasat-3 was rolled sideways frisbee-style out of the payload bay. Within minutes things started going wrong, or rather didn't start going at all. The antenna on the Leasat didn't deploy and thrusters didn't increase the satellite's spin. Something had gone drastically amiss.

It seemed unlikely, after the initial failure, that Leasat's motor would fire to boost it into the correct orbit. *Discovery's* commander, Karol Bobko, however, wasn't going to chance it, and withdrew to a safe distance. To no-one's surprise, though, the satellite remained dead. What seemed to have happened was that a switch on the satellite hadn't been tripped, as it should have been, when the satellite rolled out of the payload bay. That switch should have started an automatic sequencing timer that would have deployed the antenna, activated the thrusters and fired the booster rocket.

Below: Gardner guides the satellite towards Joseph Allen, standing on the foot restraint mounted on Discovery's manipulator arm. Allen grabs it, and then the robot arm brings satellite and astronaut over to the payload bay. The astronauts stow it in the bay manually.

Right: Dale Gardner painstakingly maneuvers Palapa B-2 into the payload bay. In the right foreground is the stinger, which the astronauts used to dock with the satellite.

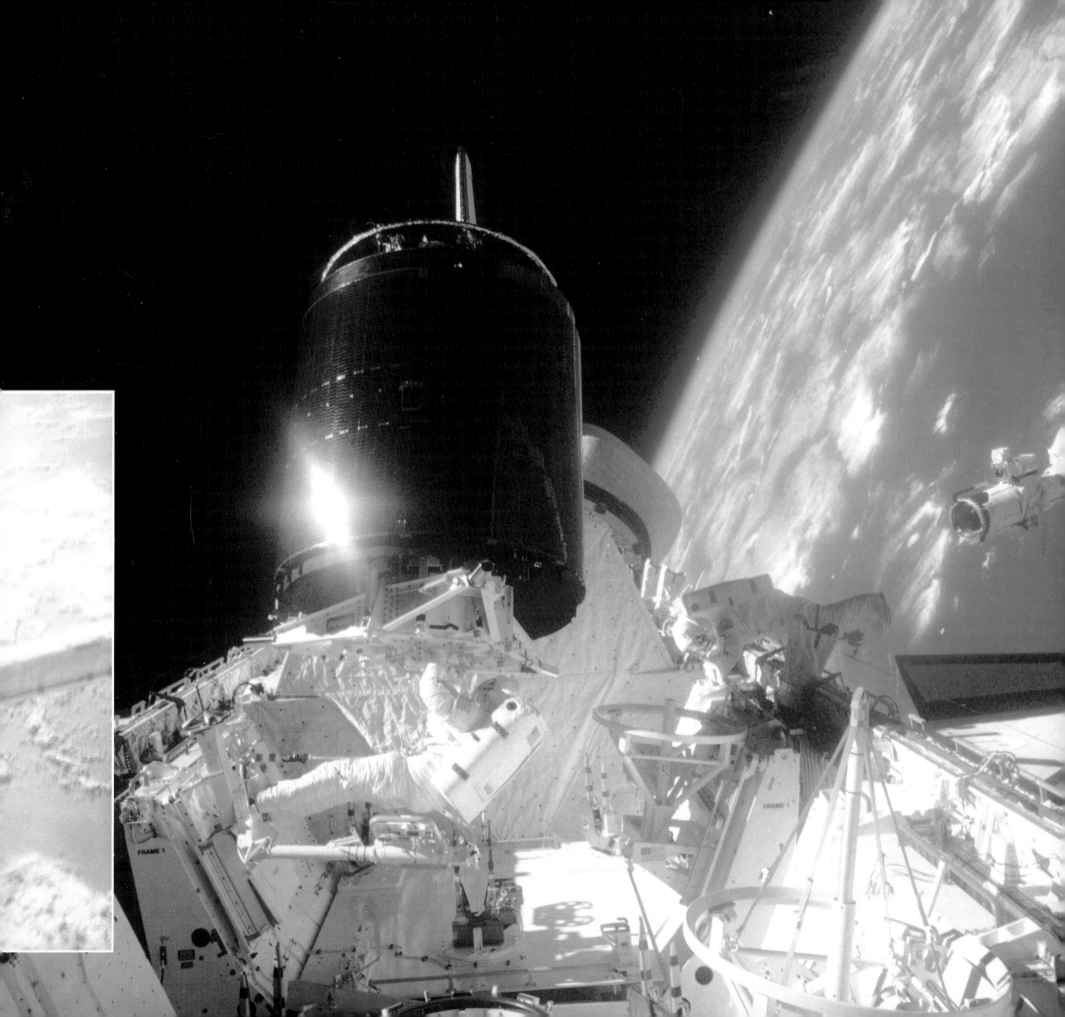

Over the next two days NASA engineers and controllers debated whether to ask the crew to attempt to activate the satellite or just make a photo reconnaissance and leave repair to a subsequent mission. Certainly there was no way they could bring the satellite back. Since the booster rocket was still intact, it still might suddenly fire. And if that happened inside the payload bay, the whole craft would go up in flames.

Eventually NASA decided to allow the crew to try and activate the time-sequencer switch with the orbiter's robot arm. But they would have to attach to it some sort of snaring device. In the finest of 'fixit' traditions, they made this from plastic notebook covers, extension rods and a metal sunshade frame. From its appearance, it was named the 'flyswatter'. On day five of the mission, mission

specialists Jeffrey Hoffman and David Griggs donned spacesuits and went into the payload bay to attach the flyswatter to the robot arm. This was the first totally unplanned and unrehearsed spacewalk of the shuttle program. It lasted just three hours.

Next day came the activation attempts, with Rhea Seddon manning the robot arm. She had only a six-minute period in which to make her attempts. Sally Ride, who had been working on snagging maneuvers in a simulator at Houston, had already talked Seddon through the various arm commands she would need. The time came, and on the third attempt, Seddon snagged the switch, but nothing happened. Nor did it when she snagged it several other times. Then the time ran out. Further attempts could have been made later, but mission control decided to call it a day.

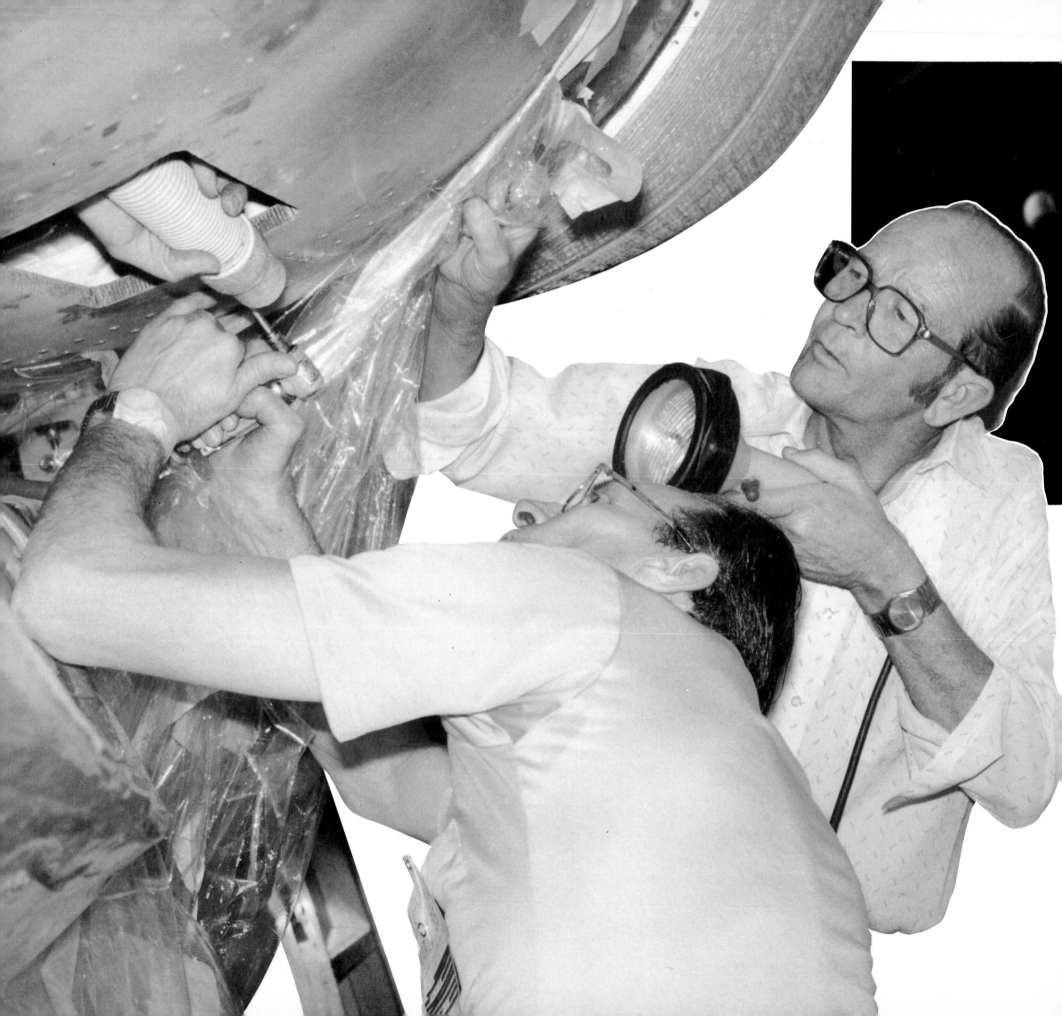

Discovery abandoned its $85 million rogue satellite — for the moment.

Hot-wiring Leasat

The loss of Leasat was a blow to NASA but a boon to the shuttle's critics, who had by no means been silenced by the spectacular achievements of 1984. And, truly, the shuttle was not proving as inexpensive a launch vehicle for satellites as had been hoped. Was it not perhaps a bit too sophisticated for the purpose? Observed one cynic: 'You don't use a Cadillac to deliver milk.' So NASA just had to go for a Leasat repair mission, and succeed. Nothing less would be acceptable.

While the details of that mission were being ironed out and rehearsed, three other shuttle missions went ahead. On 12 July 1985 the third one (51-F) set the alarm bells ringing when one of *Challenger*'s main engines shut down prematurely during launch. Fortunately the orbiter was high enough to reach orbit, though a lower one than had been planned.

The Leasat repair mission, 51-I, after two scrubbed launch attempts, began on 27 August. On board was James van Hoften veteran of the successful repair mission on Solar Max in 1984. After

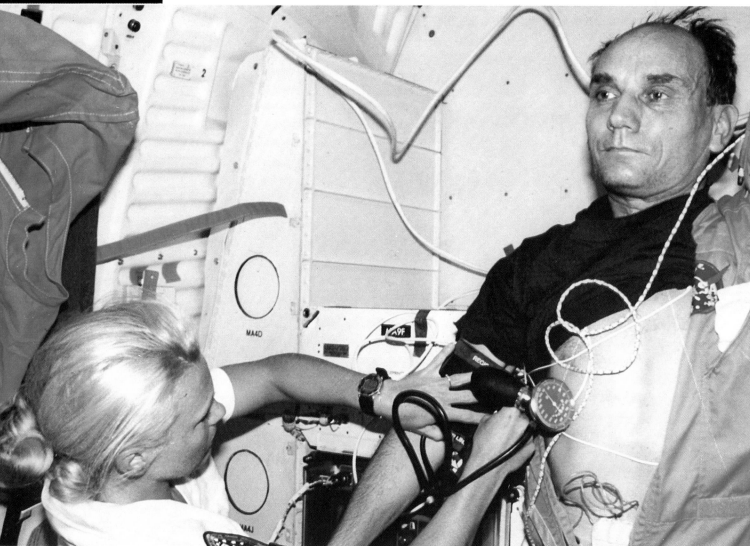

successfully deploying three satellites, *Discovery* homed in on Leasat 3, which it had left circling uselessly in orbit over five months before.

On the 31st, a spacesuited Van Hoften, nicknamed 'Ox' because of his strength, slipped into the foot restraint at the end of the robot arm. The arm carried him over to the slowly spinning Leasat. He attached a grappling bar to it and then stopped it rotating – no mean feat with a 20 foot (6 meter) long weightless but still massive cylinder. While he held onto the satellite, the robot arm carried him back to the payload bay, where fellow spacewalker and chief repairer William Fisher was waiting to 'hot-wire' the dud satellite.

Fisher skilfully fitted a new electronics box to by-pass all the circuitry and hardware that had caused the earlier failure. It would deploy the antenna, power up the satellite, and allow it to be controlled by radio command from the ground. When Fisher had finished, the satellite's antenna obediently unfolded. It was coming to life at last.

Next day Van Hoften started Leasat spinning and pushed it gently away back into orbit. The $85 million craft had been saved, added to which NASA received $55 million for its rescue efforts. *Discovery* returned to Earth in triumph, the shuttle's credibility at least partly restored. As NASA spokesman Jesse Moore said: 'This repair demonstrated the value of sending people into space, and I hope that reminds everybody of the shuttle's capabilities.'

By the time *Discovery* touched down in California, a new recruit to the shuttle fleet, orbiter *Atlantis*, was already on the launch pad at the Cape, being readied for its maiden flight in a month's time. The flight, 51-J, was to be a highly classified Department of Defense (DoD) mission. On 3 October *Atlantis* blasted off into cloudless skies after a textbook countdown. The four-day mission, during which two military communications satellites were deployed, took the shuttle to its highest-ever altitude – some 320 miles (430 km). The new orbiter performed flawlessly throughout.

Deutschland-1

On 30 October it was *Challenger's* turn to climb again into orbit. The countdown was perfect, as was the launch, precisely at noon. This mission (61-A) was a Spacelab flight, prepared and funded by West Germany. It was designated D-1 (Deutschland-1). The crew of eight, the largest ever, included three payload specialists from the European Space Agency – two from West Germany, one from the Netherlands.

In the week-long flight the crew carried out more than 70 experiments in materials science and crystal growth; in the biological and medical sciences; and in navigation and communications. One of the most interesting experiments involved the use of the vestibular (or space) sled, in which astronaut guinea pigs

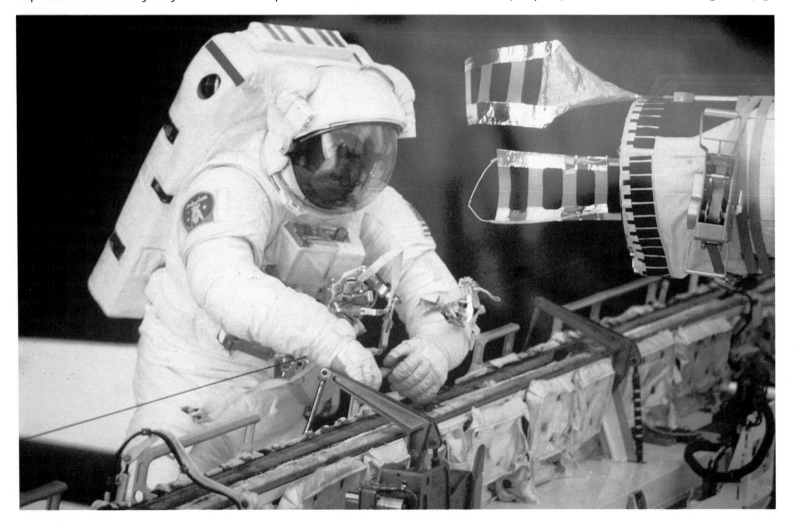

could be accelerated along a track. In the experiment they wore elaborate headgear which stimulated the eyes and ears. The objective was to gather data that might shed some light on the problem of space adaptation syndrome, commonly referred to as space sickness.

Framework for the future

From start to finish the Spacelab D-1 flight was flawless. So was the one that followed, 61-B, with orbiter *Atlantis* making its second flight. It lifted off, spectacularly at night, on 27 November. After the nominal deployment of three communications satellites, the crew prepared for the most exciting part of the mission, practicing construction techniques for future space structures,

The crew involved in the construction experiments were rookie astronauts Jerry Ross and Sherwood Spring, working in the payload bay. Assisting them from the inside was Mary Cleave, who was manning the robot arm. Ross and Spring used two different kinds of apparatus to build structures, called ACCESS (assembly concept for construction of erectable space structures) and EASE (experimental assembly of structures in EVA). In both cases they constructed frameworks from shorter aluminum beams and struts.

The spacewalkers, who had already practiced the techniques repeatedly in water simulations, made the work look child's play, assembling and dismantling the structures with ease. Television coverage of their antics made some of the most compulsive viewing of the shuttle age. On their return elated NASA officials dubbed the mission 'the best flight ever'.

Gremlins strike again, and again . . .

After three perfect missions in a row, it looked as if things were going NASA's way at last. The shuttle was delivering the goods. So it was with high hopes that they awaited the 18 December launch of *Columbia*, returning to service after a two-year refit. But the gremlins had not gone away – they had just been biding their time. Difficulties in closing out the aft compartment of the orbiter caused the launch to be postponed 24 hours. Then on the 19th the launch computer halted the countdown at T-14 seconds (14 seconds before lift-off) because of an anomaly in a SRB motor.

With Christmas fast approaching, the launch was again rescheduled, first to 4 January 1986 and then to the 6th. But on the 6th the countdown stuck at T-31 seconds when some 1500 gallons (nearly 7000 liters) of liquid oxygen propellant escaped from the fuel tank. Next day, at T-9 minutes, bad weather scrubbed the launch. In the event this was fortunate. During a routine checkover later, a 5-inch (13-cm) long sensor was discovered in an engine valve. Had the shuttle launched as scheduled, the engine might have blown up.

Heavy rains on the pad on the 10th foiled yet another attempt to launch what the Press were now dubbing 'mission impossible'. At last, on the 12th, 61-C made it into orbit. But it was a lack-luster mission, further marred when bad weather delayed *Columbia*'s return and caused it to divert to California. Meanwhile at the Cape, the knock-on effect of 61-C's delay was holding up the next mission, 51-L.

Right: An interesting first on mission 51-F in July 1985 is the provision of cola by well-known manufacturers. Ordinary pressurized cans cannot be used in zero-g, so Coke and Pepsi independently designed special containers (NASA calls them carbonated beverage dispensers) to preserve the characteristic fizziness of the drinks.

Right: In August 1985 preparations are finalized for an attempt to catch and repair the Leasat 3 satellite that refused to function after launch. Here William Fisher, mission specialist for the Leasat repair mission (51-I) is rehearsing procedures in the Houston neutral buoyancy chamber. He is working on a full-scale mock-up of the dead satellite.

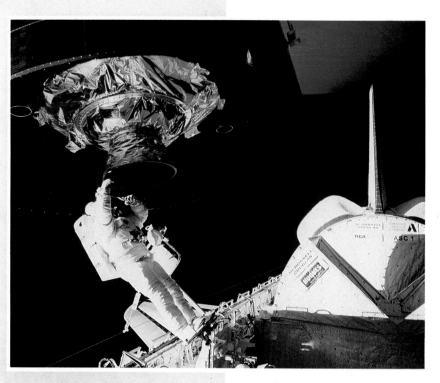

Left: Up in orbit on 31 August, James van Hoften is moved towards Leasat 3 on Discovery's manipulator arm. Later he will attach a bar to the huge satellite that can be gripped by the end effector on the arm.

Above: With Leasat 3 held by the manipulator arm, William Fisher fits a protective cover over the satellite's engine nozzle before he starts repair work on its faulty electronic components.

The Challenger Seven

Mission 51-L was rescheduled from 22 January to the 23rd, and then in turn to the 24th, 25th, 26th, 27th and finally to the 28th. Technical mishaps and again the weather were the culprits.

Tuesday 28 January dawned bright and clear over the expanse of swampland that covers the greater part of the Kennedy Space Center. But it was bitterly cold. Overnight the temperature had fallen to an unseasonable −2 C (27 F). On launch pad 39B orbiter *Challenger* stood mated to a gantry that was festooned with icicles. The countdown for its 10th and the shuttle's 25th flight was proceeding smoothly, and the weather forecast was good.

The launch team, spectators and in particular tens of thousands of schoolchildren across the nation, were waiting eagerly to see *Challenger* take to the skies. They were ill-prepared for the tragedy that would later unfold before their eyes.

At 8 o'clock local time the crew of seven astronauts, clad in their distinctive cobalt-blue flight suits, set out for the launch pad, where they rode the elevator to the orbiter cabin and strapped themselves in. In the commander's seat was Francis R. (Dick) Scobee, born in Cle Elum, Washington (1939). He had been pilot on the Solar Max repair mission (41-C) in April 1984. He once said: 'It's a real crime to be paid for a job that I have so much fun doing.' Pilot of 51-L was Michael J. (Mike) Smith, born in Beaufort, North Carolina (1945), awaiting his first shuttle flight.

Behind them on the orbiter's flight deck were mission specialists Judith A. (J.R.) Resnik and Ronald E. McNair, both poised for their second flight into space. J.R., ever smiling, a gourmet cook and classical pianist, was born in Akron, Ohio (1949). She became the second American woman to venture into space (mission 41-D, August 1984). Ron McNair was born in Lake City, South Carolina (1950).

Below: The crew of Challenger *for the 51-L mission pose for the traditional crew portrait some months before lift-off. From left to right, back row, they are: Ellison Onizuka, Christa McAuliffe, Greg Jarvis and Judy Resnik. Front row: Mike Smith, Dick Scobee and Ron McNair. Said President Reagan on the day the* Challenger Seven *died: 'They slipped the surly bonds of Earth to touch the face of God.'*

and woman in the street, she had captured the imagination.

On that launch day, 28 January 1986, millions of school-children watching on TV in classrooms across the nation, counted down the final seconds before lift-off with the crew and Kennedy launch control. At 11.38 local time *Challenger*'s three main engines roared into life, followed seconds later by those of the twin solid rocket boosters (SRBs). The explosive bolts that held down the shuttle stack were severed, and *Challenger* blasted off into the clear blue sky. Sixteen seconds into the flight *Challenger* began its accustomed graceful roll on its back. At 35 seconds the main engines throttled back, as programmed, while the shuttle passed through the zone of greatest atmospheric turbulence.

At 52 seconds mission control informed: '*Challenger*, go with throttle up.' The main engines had now throttled up to full power, and the shuttle began experiencing a condition known as 'max Q', in which it was subjected to the maximum physical stresses of the flight. At 70 seconds *Challenger*'s commander Dick Scobee confirmed: 'Roger, go with throttle up.'

The three remaining crew members were located on the orbiter mid-deck. Mission specialist Ellison S. Onizuka, born in Kealakekua, Hawaii (1946), of Japanese descent, was also aiming for his second shuttle flight. Payload specialist Gregory B. Jarvis, born in Detroit (1944) was a rookie. So was the final member of the crew, teacher S. Christa Corrigan McAuliffe, born in Boston (1948), mother of two children and a high school teacher in Concord, New Hampshire.

The 'teacher's flight'

Christa had beat off competition from over 11,000 other American teachers for a place on mission 51-L, dubbed the 'teacher's flight'. During the mission she was scheduled to conduct lessons from orbit to be broadcast live to the nation's schoolchildren, giving them a personal introduction to spaceflight they would never forget. Christa would also be the first ordinary citizen to be lofted into space, emphasizing that space flight on the shuttle was now a routine and commonplace event, open to ordinary mortals and not just to super-trained astronauts. As teacher and representative of the ordinary man

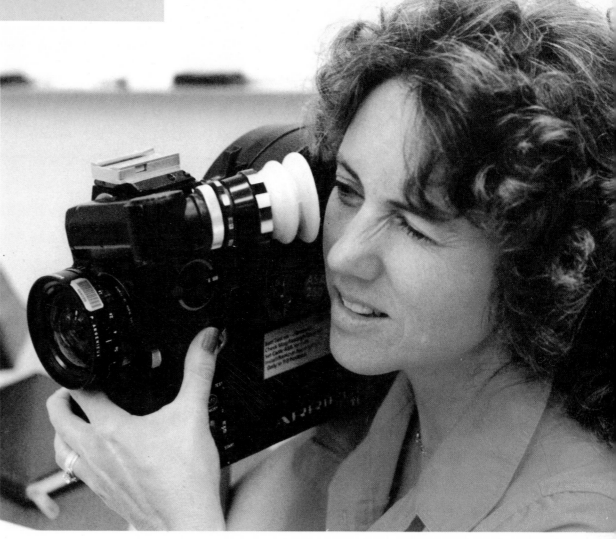

Above: Orbiter Challenger *returns from its ninth mission (the Spacelab D1 flight, 61-A) on 6 November 1985. Carrying a record crew of eight, it makes a characteristically perfect touchdown at Edwards Air Force Base in California. On its* *return to the Kennedy Space Center, it will be fitted out for its tenth and final lift-off, on mission 51-L.*

Right: 'Ordinary citizen' Christa McAuliffe practices using a movie camera during training for her flight.

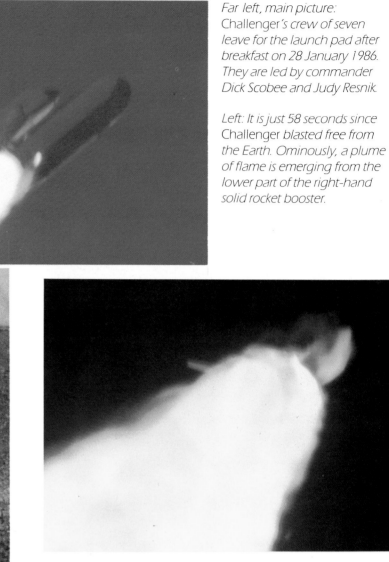

'A major malfunction'

The situation of the shuttle appeared 'nominal' — everything seemed to be going according to plan. Nothing could have been further from the truth. Just three seconds later, as onlookers followed the dense contrail of flame and smoke from the SRBs high into the sky and *Challenger* was lost to view, a mighty fireball erupted around the shuttle.

As the exhilaration of the watchers turned to horror, the voice of mission control continued reading out preprogrammed data that should have been telemetered back from the shuttle at that time: 'One minute 15 seconds. Velocity 2900 feet per second. Altitude nine nautical miles. Downrange distance seven nautical miles.' The mission control communicator was oblivious to the tragedy that had unfolded at the Cape. But then he saw the TV monitor, and the telemetry data cut-off. 'Obviously a major malfunction,' was all he could find to say.

In the skies above the Cape, the two SRBs were emerging wildly from the fireball, leaving white trails behind them. The pattern of smoke and flame resembled, as *Time* magazine later put it: 'a giant monster in the sky, its two claws reaching

blowtorch on the lower strut that attaches the SRB to the external fuel tank. At 72 seconds the strut severs. The booster pivots around the upper strut, and its nose smashes into and ruptures the external tank. A second later there is a white flash as hydrogen and oxygen leaking from the tank ignite, triggering off the fireball explosion and shattering the orbiter into pieces.

The shattered illusion

Also shattered by the *Challenger* holocaust was the illusion that space flight had become routine. As one commentator put it, the disaster 'served as a salutary reminder of the enduring hostility of space and the fundamental fragility of Man.'

Fifty-five times previously Americans had roared into space and even journeyed to the Moon, and fifty-five times they had returned safely to Earth. Speaking immediately after the disaster John Glenn said: 'We always knew there would be a day like this. We're dealing with speeds and powers and complexities we've never dealt with before.'

President Ronald Reagan at his most eloquent led the nation's mourning for the 'Challenger Seven'. At a memorial service at the Johnson Space Center, Houston, home of the astronaut corps, he recalled that 'this America was built by men and women like our seven star voyagers, who answered a call beyond duty ... Dick, Mike, Judy, El, Ron, Greg and Christa — your families and your country mourn your passing. We bid you goodbye, but we will never forget you.'

Later the President reaffirmed the nation's commitment to the manned space program. 'Sometimes when we reach for the stars,' he said, 'we fall short. But we must press on despite the pain.' Newly appointed head of the shuttle program, veteran astronaut Dick Truly concurred: 'The NASA can-do spirit is intact. We are going to get back on the track of exploring and exploiting space.'

frantically forward.' Everyone watching willed *Challenger* to emerge from the fireball too. But it didn't. Nothing could have survived the fireball explosion, as half a million gallons of liquid hydrogen and liquid oxygen in the external tank exploded.

Challenger had been ripped apart. All that was left was debris, which was still falling from the sky an hour later. Immediately afterwards aircraft, ships, submarines and divers moved into the disaster area to begin picking up the pieces of the ill-fated shuttle. But not until mid-March was the crew compartment located, containing the remains of the astronauts.

So what had happened to cause this, the worst tragedy of the Space Age? Detailed examination of photographs taken during *Challenger*'s brief tenth flight and analysis of telemetered data point to the following scenario.

Less than half a second after lift-off a puff of black smoke emerges from the lower joint of the right-hand SRB. It has been blown past the synthetic rubber O-rings that should seal the joint. Then the leak seals itself. But 59 seconds into the flight internal pressure in the booster forces gas past the O-rings again. A jet of flame spurts from the joint and plays like a

Chapter 7

Satellite Servants and Planetary Probes

E ver since the Space Age began it has always been the manned flights that have hogged the headlines and excited the imagination, and quite rightly so. But we have derived much more tangible benefits from space from the swarm of unmanned satellites that have been launched into orbit. Currently some 250-300 satellites are operational. Most of them have been launched by the United States and Russia, the remainder by ESA (the European Space Agency), Japan, China and India.

Communications, meteorology and Earth-survey are just three of the fields that have literally been revolutionized by satellite technology. When you dial an overseas call these days, it will probably travel via a communications satellite 22,300 miles (35,900 km) up in space. There, in geostationary orbit, in a 'fixed' position relative to the Earth, it relays signals from one country's communications system to another's. Weather satellites have taken most of the guesswork out of forecasting by providing hour by hour pictures of the build-up of weather systems all across the globe. No wind, no cloud, no precipitation, no depression, no cyclone can escape scrutiny by the satellites. When a geologist goes prospecting for minerals these days, he doesn't immediately, as he used to, grab for his hammer and boldly go into uncharted territories. First he consults images of the Earth's surface taken by Earth-survey satellites.

NASA satellites have carried out the pioneering work in all these fields, as well as in most other major areas of Earth resources and space science. Even its first satellite, Explorer 1, discovered something — that the Earth is girdled with belts of radiation. An outline of some of NASA's major satellite projects follows, reflecting the breadth of influence space technology has had on our everyday lives and the contributions it has made to our scientific knowledge about the Earth and the universe.

The satellites themselves have been of all shapes and sizes, from spheres a few inches across (Vanguard) to craft with antennas three times the height of Washington Monument (Radio Astronomy Explorers). Their design and instrumentation have varied greatly, depending on their function. They have all been powered, however, by panels of solar cells.

Left: False-color wizardry is applied to the Voyager images of Saturn to produce pictures like this. The technique enhances contrast between the atmospheric bands circling the planet. They consist of wind belts traveling at speeds up to 1100 mph (1800 km/h).

Right: The smallest member of NASA's launching stable is the solid-propellant rocket Scout. Launchings of this rocket generally take place at the Wallops Flight Facility on Wallops Island off the coast of Maryland. Here the Scout is being raised into its launch position in August 1964. It is carrying the satellite Explorer 20, which will probe the extreme upper atmosphere.

Far right: A Delta rocket blasts away from the Delta launch complex at Cape Canaveral in 1983. Notice the strap-on solid rocket boosters clustered around the base, which give additional thrust for the first few minutes of lift-off. This Delta is carrying a NOAA weather satellite into orbit.

Until the shuttle became operational in 1981, all NASA satellites were carried into orbit by expendable rockets, mainly Delta, Atlas-Centaur and Scout. These vehicles in fact, are still in NASA's launch stable, despite the advent of the shuttle. The Delta is the major vehicle with over 150 launchings to its credit. Continuously developed since its introduction in 1960, it stands some 116 feet (35 meters) on the launch pad. It is a three-stage rocket with nine strap-on solid boosters to augment the take-off thrust. The 131 feet (40 meter) tall Atlas–Centaur is a more powerful two-stage vehicle that can carry heavy payloads into geostationary orbit and send probes to the planets. The Scout is an all-solid four-stage rocket that can place satellites into low Earth orbit.

Comsat progress

NASA began its communications satellite (comsat) program in 1960 when it launched a satellite that inflated in orbit. An aluminized plastic balloon 100 feet (30 meters) across, it could easily be seen from Earth with the naked eye as a bright star traveling through the night sky. Appropriately named Echo 1, it was used to bounce radio signals from one ground station to another across the North American continent and also over the Atlantic.

Two years later the first true comsat was launched, named

Above: An Atlas-Centaur climbs away from Cape Canaveral in May 1983. It is NASA's heavy launcher for placing large communications satellites into geostationary orbit. This one is carrying the powerful Intelsat V, capable of handling 12,000 separate telephone conversations, plus two TV channels, plus two maritime communications links, all at the same time.

Right: Streaking like a slow-moving meteor through the Milky Way is NASA's first communications satellite Echo I, launched in August 1960. A 100 foot (30 meter) diameter balloon, it reflects radio signals between Earth stations.

Telstar. This was, in contrast to Echo, an active relay satellite. It received signals from one ground station, amplified them, and beamed them down to another. For the first time live TV pictures were transmitted between North America and Europe. The drawback of Telstar was that it was in low orbit, which meant that it had to be continuously tracked by the antennas on both sides of the Atlantic and was only receivable by both stations for 20 minutes out of each 90-minute orbit. However, it demonstrated the great potential of satellite communications.

For continuous coverage, comsats have to be placed in geostationary orbit, where they circle at the same rate as the Earth spins and therefore appear fixed in the sky. Ground stations can therefore be locked permanently on them. In 1964 NASA launched a comsat, Syncom 3, into such an orbit on the equator above the Pacific. It arrived just in time to broadcast live the opening ceremony of the Olympic Games being held that year in Tokyo, Japan.

In the same month as the launch of Syncom 3, August 1964, a group of nations throughout the world joined together to finance the development of a global telecommunications system founded on satellites. It was named Intelsat (International Telecommunications Satellite Organization). Its first satellite, Intelsat I, or 'Early Bird', was placed in orbit over the Atlantic in 1965. With a capacity of 240 voice circuits, it alone increased the existing transatlantic telephone capacity by 50 per cent. Two further satellites were launched in 1967 over the Pacific and Indian Oceans. There, spaced at equal distances around the Earth, they provided the first global communications links.

Today more than 110 countries belong to Intelsat, which uses up

Above: Seeing is believing. The first Intelsat satellite (1965), is shown next to a Hughes technician. Both are dwarfed by the latest member of the Intelsat series, Intelsat VI, which is depicted here life size!

Left: Another very powerful communications satellite, being readied for launch at the Kennedy Space Center, called a tracking and data relay satellite (TDRS).

Right: Government and business organizations now routinely use satellite communications. America's Earth-resources data center EROS, near Sioux Falls, South Dakota, receives Landsat and other Earth-sensing data via satellite from the Goddard Space Center near Washington.

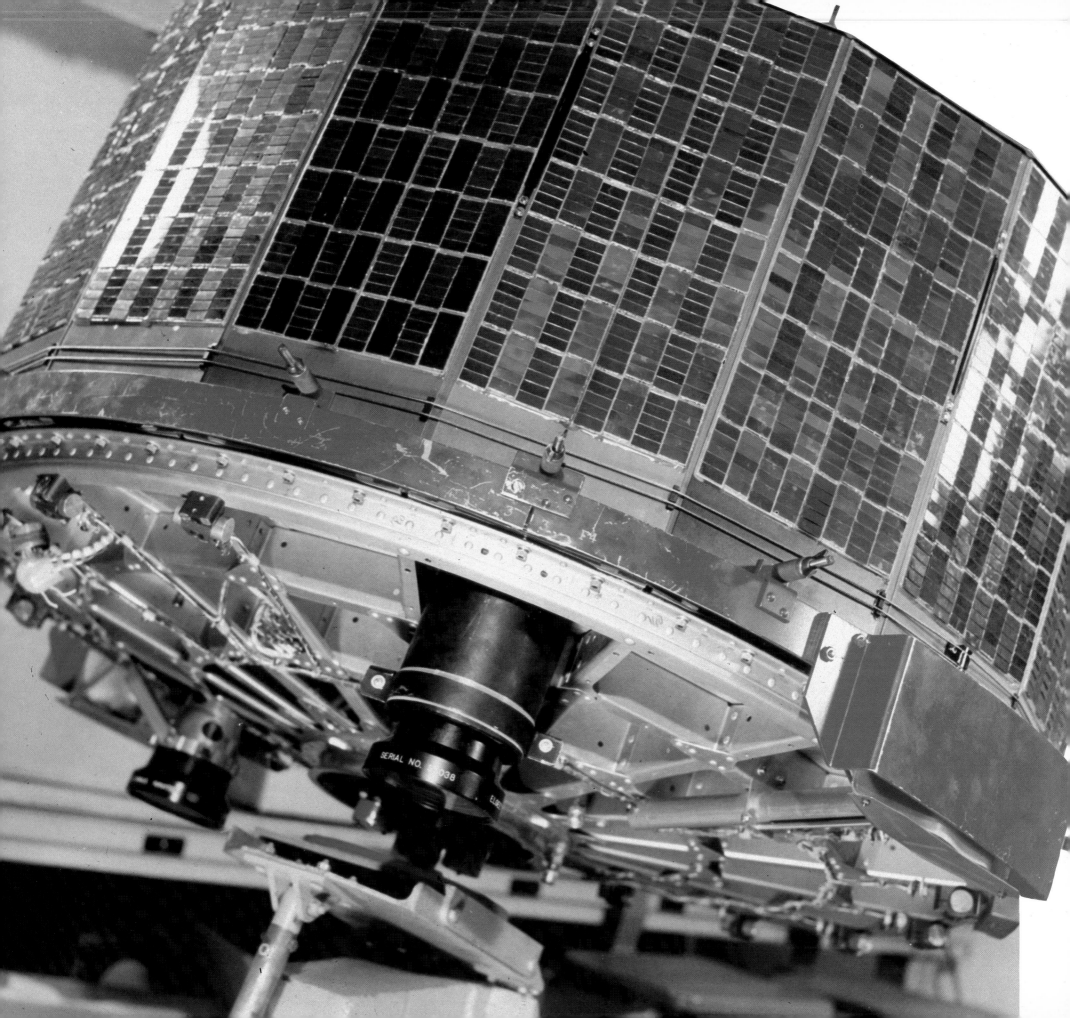

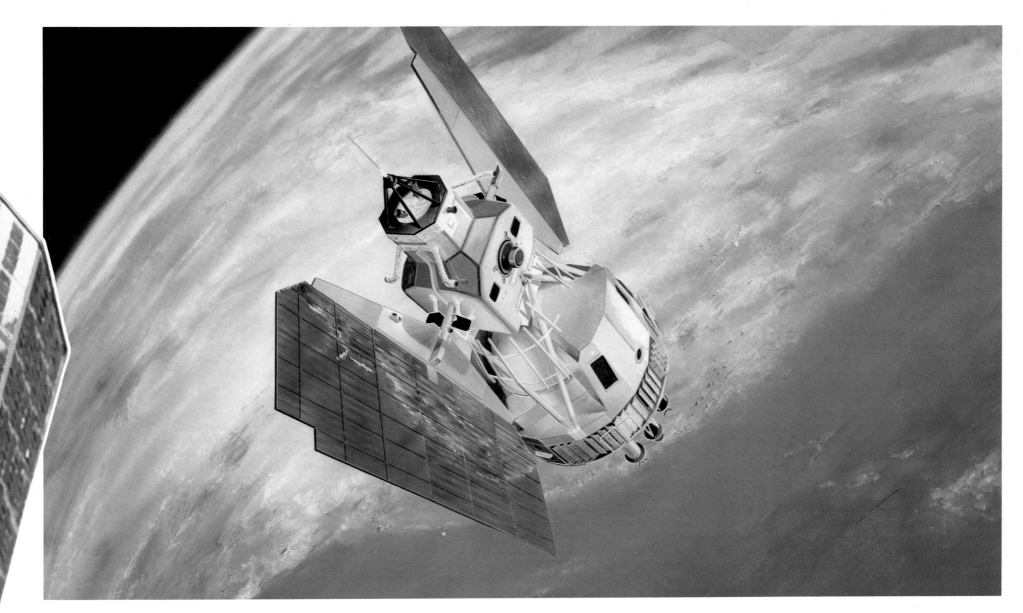

to 20 satellites to relay telephone, telex, television, data, and
facsimile signals between all the continents except Antarctica. The
most powerful comsats in service at present are the Intelsat V and
VI. The former can relay 12,000 two-way telephone conversations
at once, as well as two color TV channels. Intelsat VI can relay over
30,000 telephone conversations. Both are huge in size. Intelsat V is
powered by an array of solar cells measuring some 52 feet (16
meters) from tip to tip. The cylindrical Intelsat VI stands some 40
feet (12 meters) high and is 12 feet (3.6 meters) across.

Keeping a weather eye
NASA also began its weather satellite program in 1960, with the
launch of Tiros 1. 'Tiros' stands for 'television and infrared
observation satellite'. This low-orbiting craft took cloud-cover
pictures of the mid-latitudes of the Earth and only in the daytime.
Tiros 2, launched later in the year, gave wider coverage and had
infrared sensors so that it could send back pictures at nighttime as
well. From that time there has always been at least one satellite
aloft reporting on the weather. The Tiros family of spacecraft

improved with each successive launch. In 1963 Tiros 8 inaugurated
the APT (automatic picture transmission) system, which allowed
weather forecasters anywhere to receive local cloud-cover
pictures and other data directly via simple antennas.

Tiros 10, launched in 1965, was the first satellite funded by the
US Weather Service. Subsequent satellites were designated ESSA
when the Weather Service became part of the Environment
Science Services Administration. When ESSA in turn was absorbed
into the National Oceanographic and Atmospheric Administration, the satellites became designated NOAA.

Another major weather satellite was Nimbus. It was a very
distinctive-looking spacecraft with moth-like solar wings. It served
as a flying testbed for new instruments that were later
incorporated into Tiros, as well as ESSA and NOAA operational
weather satellites.

The current NOAA satellites are a far cry from Tiros 1. They are of
a design known as advanced Tiros-N. Nearly 3800 pounds (1700
kg) in weight and costing some $45 million to build, they are highly
sophisticated robots. They circle in polar orbit some 550 miles (880

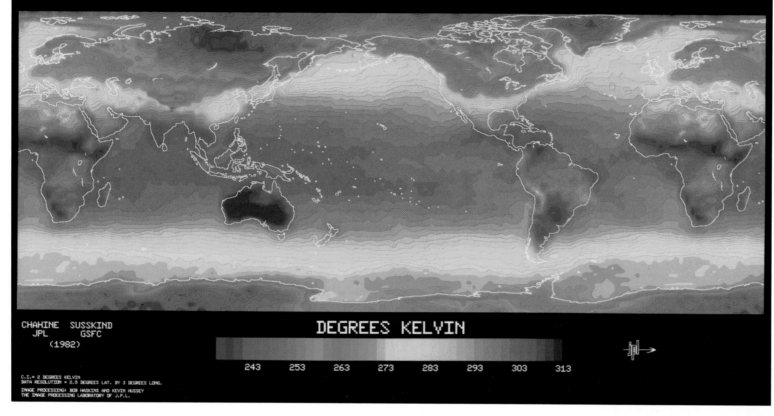

MEAN SURFACE TEMPERATURE FOR JANUARY 1979
FROM HIRS 2 AND MSU DATA

CHAHINE   SUSSKIND
   JPL   GSFC
      (1982)

DEGREES KELVIN

243   253   263   273   283   293   303   313

C.I.= 2 DEGREES KELVIN
DATA RESOLUTION = 2.5 DEGREES LAT. BY 3 DEGREES LONG.
IMAGE PROCESSING: BOB HASKINS AND KEVIN HUSSEY
THE IMAGE PROCESSING LABORATORY OF J.P.L.

Right: A highly advanced weather satellite called Tiros goes into orbit in 1978 and returns superior atmospheric data. Computer-processing techniques applied to the data enables meteorologists to produce cloud cover pictures like this one, which shows a classic cyclonic system developing in the North Atlantic and sweeping towards Europe.

Below: This is the normal type of image of cloud cover received from polar-orbiting NOAA satellites. This one is acquired from NOAA-5 in 1978. It shows a near cloudless central and eastern United States, but dense cloud banks, however, are forming over the Atlantic and Caribbean.

Above: The advanced sensors on the latest Tiros/NOAA polar orbiting satellites continuously monitor the temperature of the Earth's surface. This global temperature profile is prepared from data received from the pioneering Tiros N in 1979. Note that the hottest regions are in Africa, Australia and South America, south of the equator.

km) high and orbit every 100 minutes as the Earth spins beneath them. Their instruments picture cloud cover, monitor temperatures at the surface and at different levels in the atmosphere. They also relay environmental data from unmanned weather stations, buoys and balloons.

Interestingly some of them also carry search and rescue equipment that relays distress signals from wrecked ships and boats and crashed planes. They form part of a search and rescue satellite-aided (SARSAT) emergency service operated in conjunction with Cospas satellites of the Soviet Union. Between 1982 when it became operational and early 1986, more than 2000 distress calls had been relayed through the SARSAT network. Though most were false alarms, some were for real. And it is reckoned that the network saved over 350 lives over that period.

Earthscanning

Photography of the Earth from orbit has been one of the most fascinating features of the Space Age. It began on the Gemini missions in the 1960s and continued through Apollo and Skylab into the present shuttle era. Ordinary photographs from space present us with extraordinary views of our colorful living planet, but reveal nothing our eyes cannot see. Photography in light of different wavelengths, however, throws up details of the landscape that are invisible to human eyes.

To exploit this phenomenon, NASA launched a series of Earth-resources technology satellites, beginning in 1972. The first was named ERTS-1, later renamed Landsat 1. This had a similar

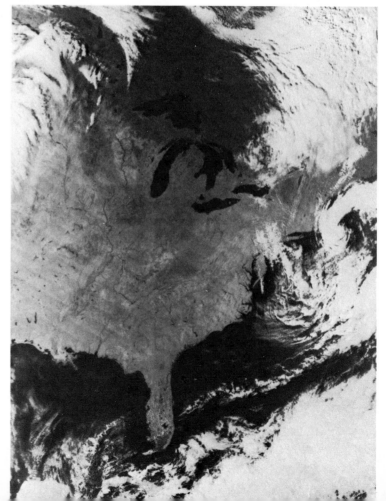

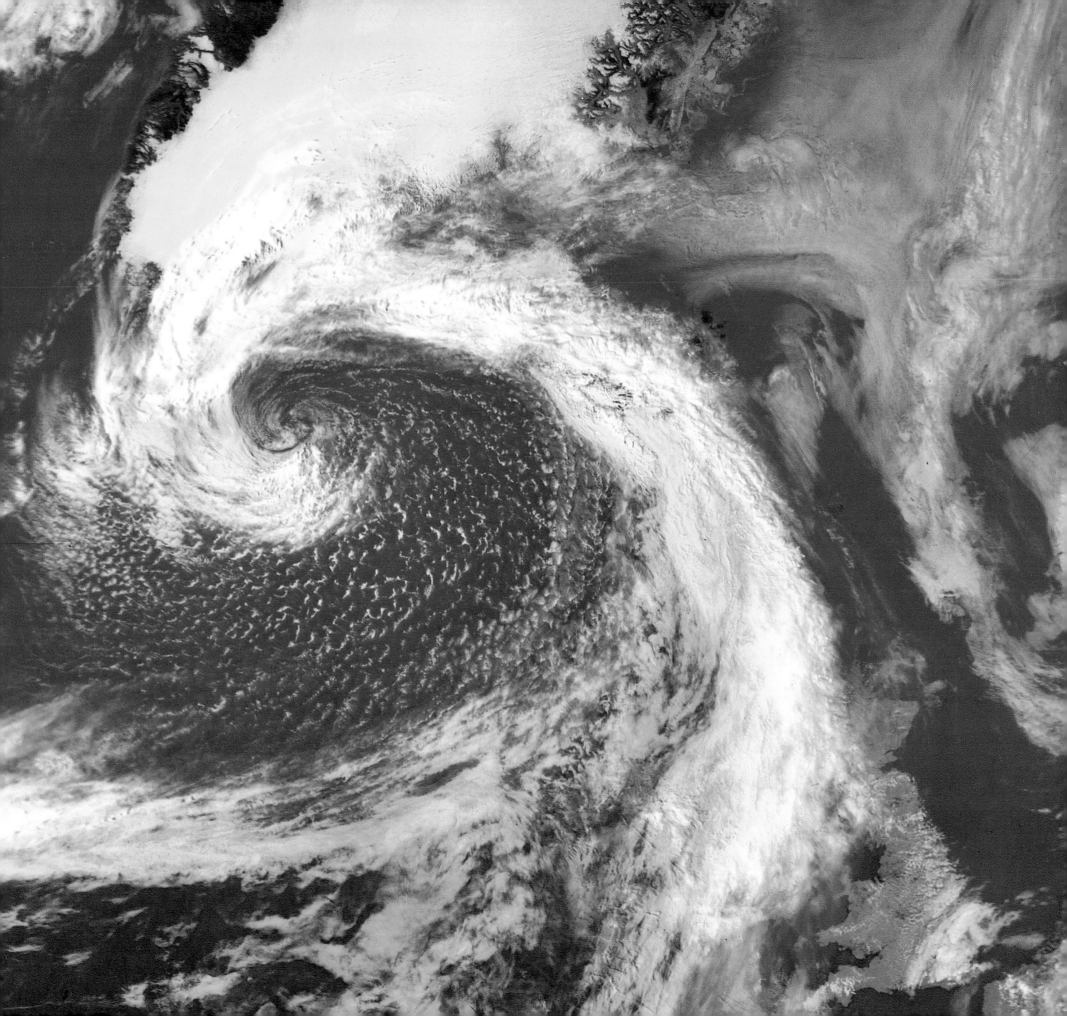

Right: This is one of the latest kind of weather satellites known as GOES (geostationary operational environmental satellite). Unlike the NOAA satellites, GOES craft operate in geostationary fixed orbit and thus view the same part of the globe all the time. GOES satellites over the Atlantic and Pacific provide cloud cover pictures of the Americas every half-hour.

Left: This simulated natural color image of the Americas is produced from GOES data received in 1979. It shows three storm centers in the Atlantic and a massive one in the Caribbean, which is actually the destructive hurricane David.

moth-like design to the Nimbus series of weather satellites, as did Landsats 2 (1975) and 3 (1978). Landsats 4 (1982) and 5 (1984) were much more advanced, and the pictures they took could resolve smaller details in the landscape. Like the nation's weather satellites, Landsats are now under NOAA management.

The latest Landsat spacecraft have two picture-taking systems called the multispectral scanner (MSS) and the thematic mapper (TM). They do not take actual photographs, but form electronic images of the landscape by means of a scanning technique. They incorporate oscillating mirrors that direct the light reflected from the ground through filters into detectors in various visible and invisible light wavebands. The intensity of light in each waveband is converted into digital electronic data and transmitted to a ground station. The data is then computer-processed into a visible image, covering a not quite square area of 115 × 115 miles (185 × 185 km).

The computer-processing can manipulate the data in many ways to give a variety of false-color images. The colors can be selected to emphasize particular types of ground features. This is possible because each feature reflects different wavelengths of light in different ways. We say it has a different 'spectral signature'. The Landsat spacecraft can recognize these signatures, which are invisible in ordinary light.

It is almost impossible to list the number of fields of study that have benefited from Landsat images and data. Geographers and geologists have perhaps benefited the most. The Landsat images

Far left: Here in the EROS Data Center at Sioux Falls, South Dakota, a Landsat user is making a digital analysis of an image of the island of Hawaii. EROS has a catalog of millions of Landsat images, available for purchase by people throughout the world. Many countries also have their own remote-sensing centers that supply Landsat images.

Left: This Landsat 4 thematic mapper (TM) image of Los Angeles is acquired in November 1982, four months after the satellite's launch. It clearly shows the grid pattern of the streets and the expressways through and to the north of the city.

On pages 188 and 189: The city of Detroit is located at the top of this Landsat TM image. The urban areas show up as blue-gray, the vegetation in the countryside as red. Across the Detroit River is the Canadian town of Windsor. The body of water is Lake Erie.

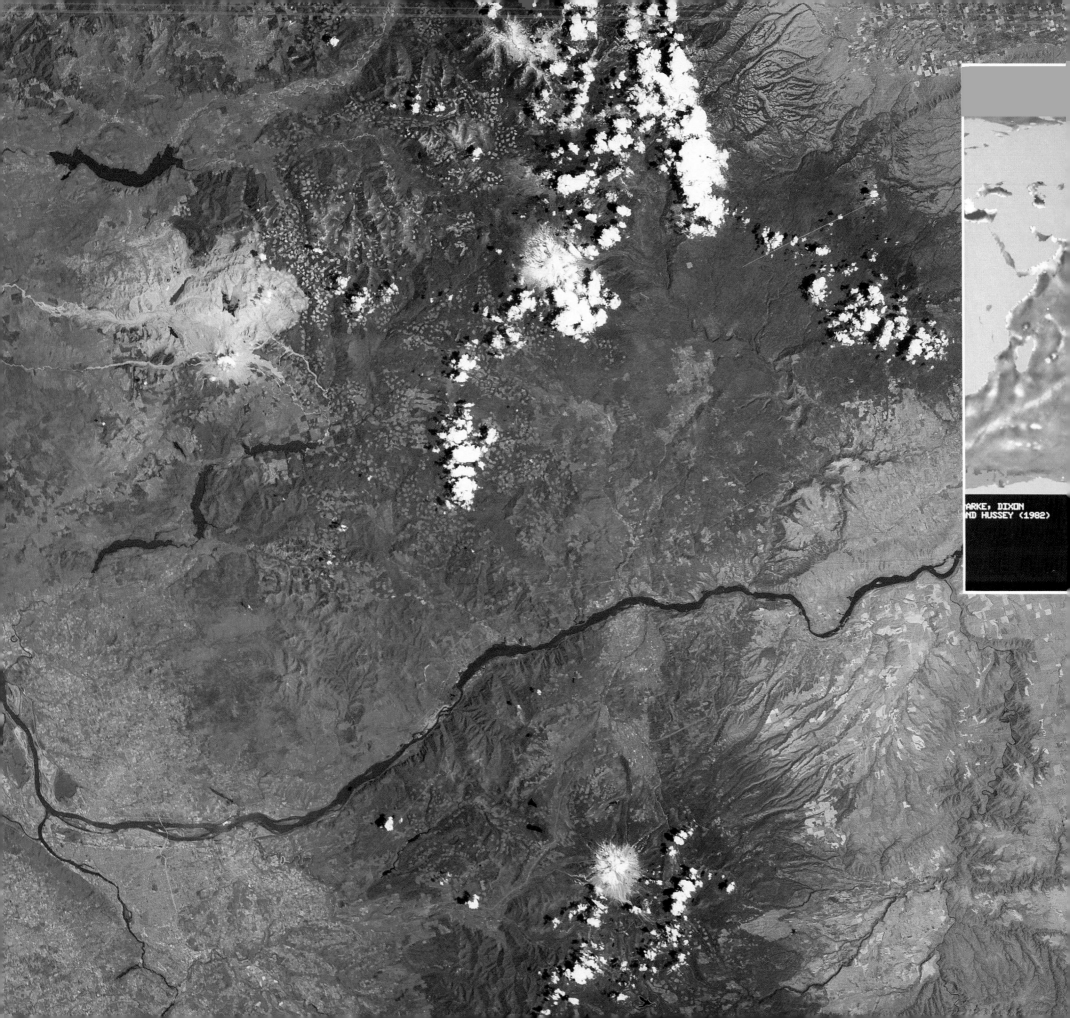

CLARKE, DIXON
AND HUSSEY (1982)

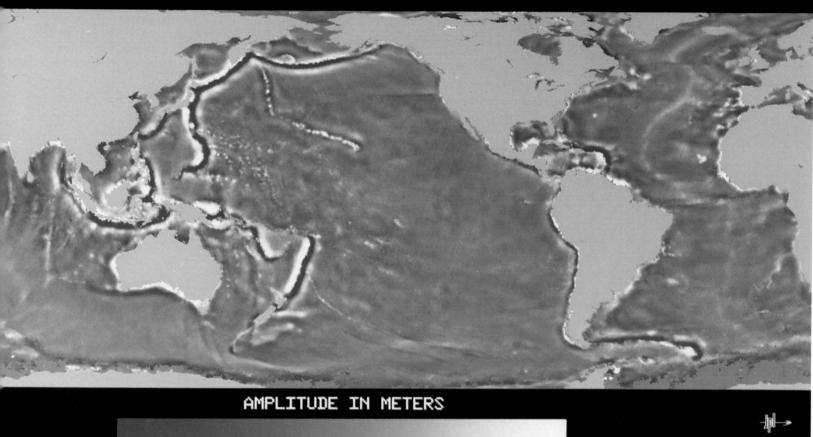

SEA SURFACE HEIGHT ANOMALIES FROM SEASAT RADAR ALTIMETER
JULY 7 – OCTOBER 10, 1978

AMPLITUDE IN METERS

-5   0   +5

Left: This relief map of the world's oceans is prepared from radar altimeter data returned by the short-lived Seasat satellite, launched in 1978. It shows unbelievably that the ocean surface dips down in places where there are submarine troughs and rises where there are ridges. So in effect the picture is also a relief map of the ocean floor!

Below: This artist's impression shows the sixth in the Explorer series of NASA scientific satellites in Earth orbit. Launched in August 1959 it sends back further information about the Van Allen radiation belts and the dust particles, or micro-meteoroids that permeate interplanetary space. It also returns a crude photograph of the Pacific.

have greatly simplified the work of cartographers, the people who make maps, especially in remote regions of the world hitherto inaccessible. Since Landsat views the world from such a lofty vantage point, geologists are able to trace large-scale features of the Earth's surface as never before. By using subtle false-color imaging they can also detect rock formations that hint of underground mineral deposits. Major deposits of copper in Chile, tin in Bolivia and oil in the Sudan have already been located by such techniques.

In agriculture, Landsat images have helped farmers in crop forecasting and planning, as well as spotting disease. Diseased crops can be identified because they have a different spectral signature from healthy ones. The different spectral signatures of ground features within cities aids urban planners, who find it impossible by conventional means to check on city growth. Landsat images also assist the study and management of forests and water resources, and provide a method of monitoring floods, beach erosion, water pollution and other environmental problems.

Left: The devastation caused by the eruption of Mount St Helens in May 1980 is graphically illustrated in this Landsat multispectral scanner image taken 14 months later. Gray ash can be seen covering the countryside for miles around, and the rivers leading from it are choked with ash too. The peak to the right of Mount St Helens is Mount Adams.

Orbiting observatories
Since its inception, NASA has placed as much emphasis on space science, that is, gathering information about space; as on space

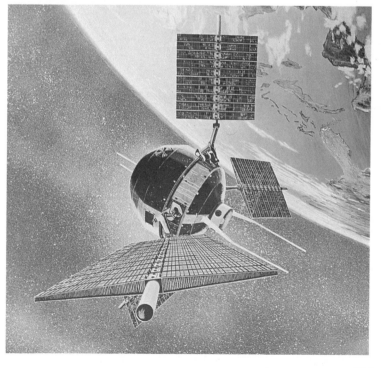

ETA CARINAE
EINSTEIN OBSERVATORY
300 ARC-SECS: ⊢———┥

applications, that is, utilizing the space environment for the direct benefit of mankind. They have launched hundreds of scientific satellites into orbit since Explorer 1 opened the American space effort on the last day of January 1958. It was the first of over 60 satellites in the Explorer program. Most of them were small and carried out investigations in three major areas: geophysics, the Earth's atmosphere and astronomy.

The geophysical Explorers, of which Explorer 1 was the first, investigated the Earth's magnetic field and the way it interacts with the solar wind – the stream of particles that issues from the Sun. They discovered and mapped the extent of the Earth-girdling Van Allen radiation belts, and found that the Earth's magnetosphere – the magnetic 'bubble' around the Earth – is shaped like a giant teardrop because of the action of the solar wind. The atmospheric Explorers (beginning with Explorer 8, 1960) discovered that the size and density of the atmosphere fluctuate. It swells by day and contracts by night. Its size is also affected by activities on the Sun, such as solar flares and sunspots.

Explorer 11 (1961), which carried gamma-ray detectors, was the first of the astronomy Explorer series of satellites. These took advantage of the above-atmosphere location to view the universe at wavelengths that cannot be studied from the ground because they are absorbed by the atmosphere. Explorer 42 (1970) probed the universe at X-ray wavelengths, which again cannot be

detected from Earth. Unusually, it was launched by Italy from the San Marco launch platform off the coast of Kenya. Since it was launched on Kenya's independence day, it was named 'Uhuru', which is Swahili for 'freedom'. Uhuru showed the X-ray sky to be far brighter than expected, and it pioneered what has now become one of the most important branches of astronomy.

Later, orbiting astronomical observatories such as Copernicus (OAO-3, 1972) and high-energy astronomical observatories such as Einstein (HEAO-2, 1978) explored the X-ray universe in more detail, finding possible evidence of the most bewildering of cosmic phenomena, black holes. These baffling objects, postulated in theory but never directly observed, are thought to be the final death throes of massive stars that have collapsed upon themselves so catastrophically that they have virtually crushed themselves out of existence. What remains are regions of space with such stupendously high gravity that nothing, not even light, can escape from them. Thus they are abysmally black.

The stars also give out light in other wavelengths that we cannot properly study from Earth – in the ultraviolet and the infrared. The ultraviolet gets absorbed by the ozone layer in the atmosphere; most of the infrared is absorbed by the water vapor there. The first major ultraviolet astronomy satellite was the international ultraviolet explorer (IUE, 1978) a joint project of NASA and the European Space Agency. Ominously, IUE data suggested that a

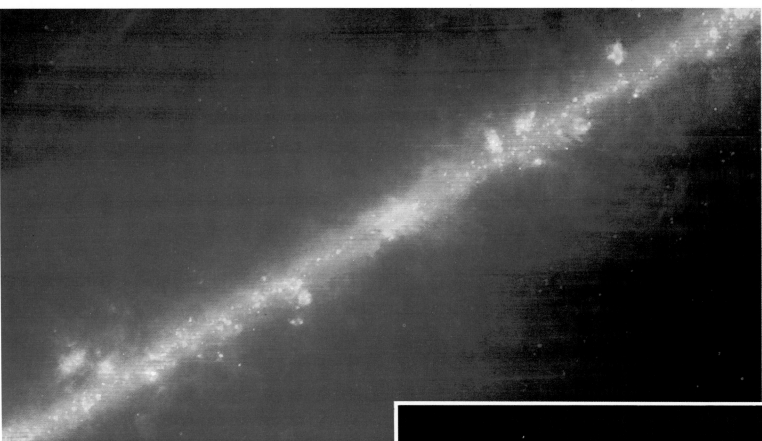

black hole with the mass of 1000 solar systems could be lurking in the center of our own galaxy.

Another outstanding international project organized by NASA, the Netherlands and Britain made the first comprehensive study of space in the infrared in 1983. Known as the infrared astronomy satellite (IRAS), it featured a telescope cooled by liquid helium at a temperature only a few degrees above absolute zero (-273°C). It was thus able to detect the infrared, or heat radiation from even the coolest stars and interstellar gas and dust clouds. It cataloged more than 200,000 infrared sources, spotted stars being born, and detected clouds of solid material around the bright stars Vega and Fomalhaut that could well be new solar systems in the making.

Probes to the Moon

NASA had scarcely mastered the art of launching satellites into orbit around the Earth, when it began to tackle the next space challenge, the launching of spacecraft, or probes, into outer space. For this to happen, the probes would have to be boosted to speeds in excess of the Earth's escape velocity of 25,000 mph (40,000 km/h) so that they could escape completely from the clutches of Earth's gravity.

You need no guesses that the target for the first shots into outer space was the Moon, our nearest cosmic neighbor, a 'mere'

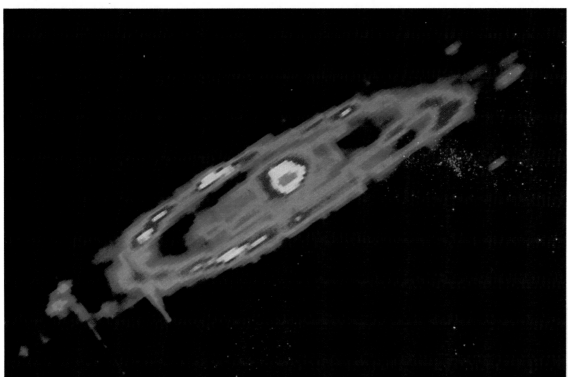

quarter of a million miles away. NASA made four attempts to reach the Moon with Pioneer probes during 1958, but none of them even achieved escape velocity. Considering the relatively low power of NASA's launching rockets at the time, perhaps this was not surprising.

It was inevitable that the Soviet Union, with its much more powerful launching rockets, would beat them to it; and they did. In January 1959 their Luna 1 probe flew within 3000 miles (5000 km) of the Moon. In the following September Luna 2 crash-landed on the Moon, becoming the first man-made object to reach another world. A month later Luna 3 orbited around the far side of the Moon and returned the first ever pictures of its surface.

NASA at the time had no immediate answer to these Russian advances apart from achieving the satisfaction of a fairly near miss with Pioneer 4 in March 1959. Then came a series of disasters with Pioneer/Orbiter probes attempting a lunar orbit. There was little initial luck, either with the Ranger program that followed. Ranger

Above: This image shows IRAS's view of the famous Andromeda galaxy, a spiral galaxy like our own and some 2 million light-years away in the constellation of Andromeda.

Right: The Hubble space telescope is poised to revolutionize optical astronomy when it becomes fully operational. It peers into the depths of space with a 7.8 foot (2.4 meter) diameter reflecting telescope. Because of its location above the atmosphere, it should be able to see seven times farther into space than we can from the ground. It transmits its data to the Science Institute in Baltimore via a tracking and data relay satellite (TDRS), a ground station at White Sands and NASA's Goddard Space Center.

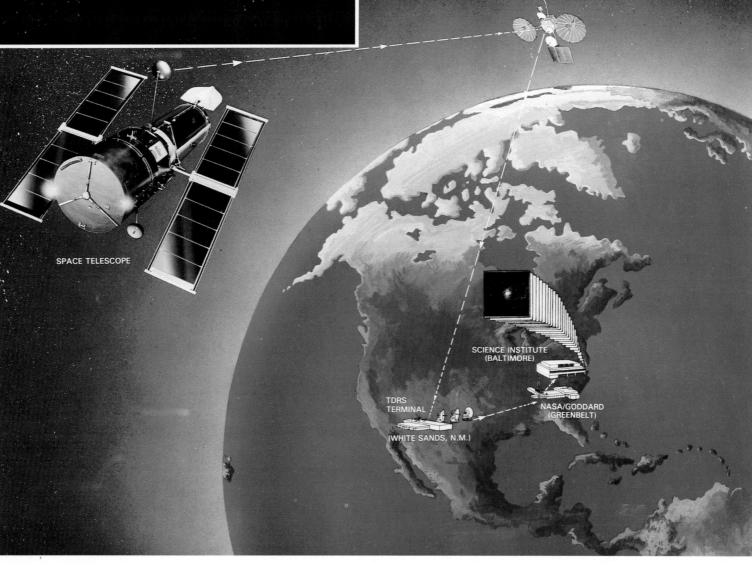

TDRS

SPACE TELESCOPE

SCIENCE INSTITUTE
(BALTIMORE)

TDRS
TERMINAL

NASA/GODDARD
(GREENBELT)

(WHITE SANDS, N.M.)

craft were intended to transmit television pictures of the surface before crash-landing. The early ones didn't reach the Moon or crashed before taking pictures.

Not until July 1964 was success achieved when Ranger 7 transmitted over 4000 high-resolution pictures before impact in Mare Nubium, the Sea of Clouds. So impressed was the International Astronomical Union with the pictures that it renamed the region Ranger photographed Mare Cognitum, the Known Sea. For NASA this was the breakthrough, and heralded the beginning of a determined and highly successful assault on the Moon that was to culminate with the manned lunar landings.

In 1965 there were two more Ranger missions, and in 1966 came the start of the Surveyor and Lunar Orbiter programs. Surveyor was a three-legged soft-landing craft used to test the feasibility of landing vehicles on the Moon, investigate the nature of the soil and take close-up pictures of the surface. Surveyor 1 alone over a six-week period in 1966 sent back more than 11,000 high-quality pictures. Lunar Orbiter craft, as the name implies, photographed the Moon from orbit and returned pictures to Earth. One of the main objectives of the Lunar Orbiter program was to reconnoiter possible sites for the Apollo landings. In doing so they photographed nearly the whole of the lunar surface, including the far side.

Exploring the inner planets

Our nearest planetary neighbors in space are Venus and Mars, some 26 million and 35 million miles (42 million and 56 million km) distant respectively at their closest approaches to Earth. Sending probes even to these 'near' neighbors presented formidable problems to the US and Russian space scientists in the early days, when propulsion, communications and tracking technology and computing facilities were primitive and unreliable compared with what they are today.

The first attempts to reach Venus and Mars began in 1960 and for two years neither Russia nor the United States had any success. Then in December 1962 the US probe Mariner 2 flew within 22,000 miles (35,000 km) of Venus and reported, astonishingly, that the temperature of its atmosphere was in excess of 425°C (800°F) – hot enough to melt lead! The atmosphere also appeared to be very much denser than that on Earth.

NASA also had the first success in exploring Mars, with Mariner 4, in 1965. The probe sent back a few low-quality pictures from an altitude of about 6000 miles (10,000 km) that revealed a heavily cratered lunar-like landscape. Another significant observation was that Mars had traces of an atmosphere of carbon dioxide at a pressure only about one-hundredth atmospheric pressure on Earth.

Right: The lunar probe Lunar Orbiter 5 took this excellent picture of the Moon's Alpine Valley in 1967 as part of its mission to map the Moon and investigate possible Apollo landing sites. The straight lines in the picture result from the Orbiter's scanning system.

Left: The 210 foot (64 meter) diameter dish antenna at Goldstone tracking station in California, located in the desert north-west of Los Angeles. It forms a key link in NASA's Deep Space Network, which communicates with probes traveling into the depths of the solar system.

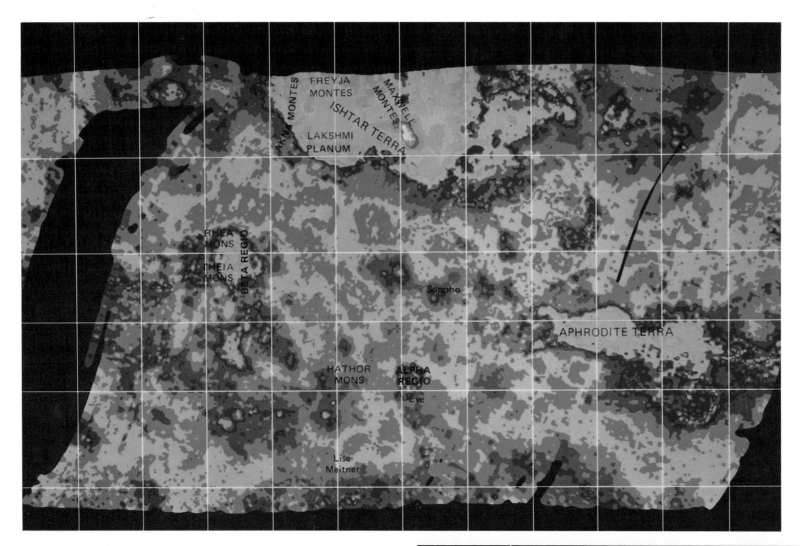

Left: This picture of the topography of Venus has been built up from data acquired from radar scans by radio telescopes from Earth and by orbiting probes such as Pioneer-Venus (1978). Radar beams can penetrate the exceeedingly dense carbon dioxide atmosphere and record ground features from echoes. The scans show two highland areas on Venus, Ishtar Terra and Aphrodite Terra. The of the planet is fairly flat.

Right: Cloud patterns show up in the atmosphere of Venus after computer enhancement of data returned by the two-planet probe Mariner 10 in February 1974. From Earth, all we see of Venus is featureless white cloud.

Below: In March 1974 Mariner 10 encountered Mercury, the planet nearest to the Sun, and took this picture of the planet's northern limb. The landscape is very reminiscent of the Moon's rugged highland regions.

With the trail-blazing missions of Mariners 2 and 4, NASA established a pre-eminence in planetary exploration that it has never lost. Its strength then, as now, lay in the sophistication of its electronics and instrumentation, forced upon it by its initial weakness in propulsion technology.

Venus again came under scrutiny in 1967 when Mariner 5 confirmed the high surface temperature and established that its atmospheric pressure is 100 times the Earth's. Mariner 10 flew by Venus in 1974 on its way to visit Mercury, taking ultraviolet pictures that showed the circulation of clouds in the thick atmosphere. It actually used the gravity of Venus to redirect it into a trajectory toward Mercury. This was the first time such a gravity-assist, or 'sling-shot', method had been used. Mariner 10 sent back the first ever pictures of the Sun-baked surface of Mercury, which proved to be highly cratered like the Moon but with no lunar-type maria (seas).

Exploration of Mars continued in 1969 with a two-pronged assault by Mariners 6 and 7. Two years later Mariner 9 went into Martian orbit and systematically mapped virtually the whole of the surface of the planet. It pictured amazing topography, including massive volcanoes, one of which, Olympus Mons, is more than three times the height of Everest and some 300 miles (500 km) wide at the base.

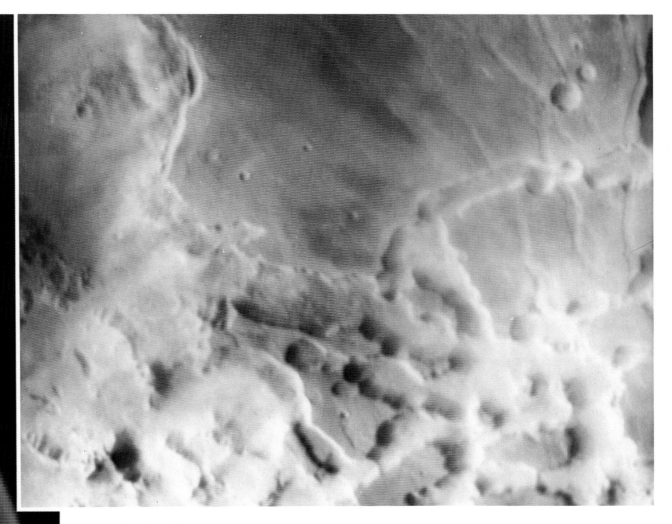

Looking for life

Another remarkable feature of the Martian landscape revealed was a huge Martian canyon, that scarred the surface for 3000 miles (5000 km). It was named Valles Marineris after the spacecraft that discovered it. Even more exciting to scientists was the evidence of what appeared to be ancient water channels, which suggested that Mars might have had a more moderate climate sometime in the distant past. But was there evidence of Martian life or of the 'canals' that astronomers have reported seeing through telescopes? There was not. Mars appeared to be a lifeless barren planet.

Some scientists, however, reckoned there might just be a chance that Mars supported primitive life forms that had adapted to the harsh Martian environment, like the lichens that exist in the Antarctic here on Earth. Or the evidence of ancient life forms might exist in organic or fossilized traces in the Martian soil. There was one way in which this matter might be resolved and that was to land a probe on Mars and test the soil for signs of life. This task became one of the major objectives of NASA's Viking mission.

In fact two identical Viking probes were despatched to Mars, both arriving in 1976. The Viking spacecraft consisted of two parts. One, the orbiter, remained in Martian orbit; while the other, the lander, descended to the surface. The Viking landers set down in plains regions known as Chryse (Viking 1) and Utopia (Viking 2).

Left: As the Viking 1 spacecraft approached Mars in 1976, it took this picture, imaged in natural color (north is towards the upper left). The most prominent features are dark circles that mark the location of four massive volcanoes. Biggest is the one on the left of the row of three, known as Olympus Mons.

Above: The rugged canyon lands of Mars are pictured here in the early morning, shrouded with mist. This region is known as the Labyrinth of the Night.

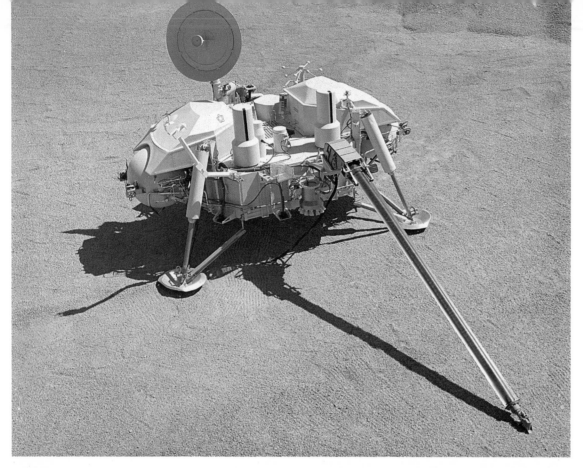

Right: This great gash in the surface forms part of the enormous canyon of Mars, called Valles Marineris. It girdles the planet for some 3000 miles (5000 km), and in places is as wide as 250 miles (400 km) and as deep as 5 miles (7 km).

Left: A replica of the Viking lander that soft-landed on Mars. Protruding in the foreground is the retractable arm used to scoop up soil and deliver it into the lander's automated biological laboratory.

Below: The retractable arm is seen on the right of this picture, which the Viking 1 lander takes from the surface of Mars. The Martian landscape proves to be a rusty red color at both landing locations.

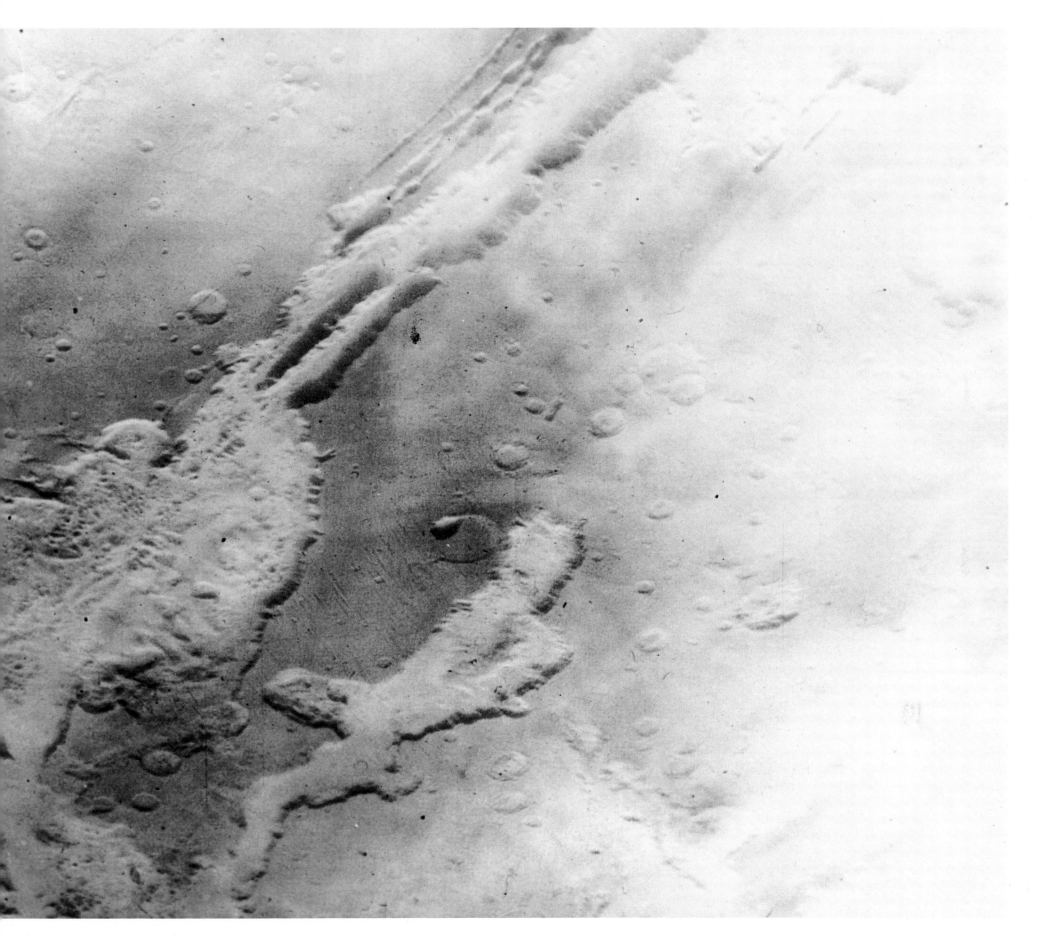

Both the orbiters and landers took remarkable color pictures of the landscape, an orange-brown rusty color almost everywhere. The close-up pictures of the two landing sites were very similar, showing dark, probably volcanic rocks scattered over soil, with dunes of finer material collected here and there.

The landers had meteorology booms to sample the weather, and a mechanical arm to scoop up soil and drop it into an automated biological laboratory. In the laboratory several different methods were used to test for the presence of organic matter that might hint of life. Vigorous reactions certainly occurred during the tests, but could not be attributed to the presence of organic matter. Probably these were caused by peculiar Martian chemistry. The tests didn't prove that there was no life on Mars, but they made it less likely.

Beyond the asteroid belt

In the solar system there is a huge 'gap' between Mars, a relatively close neighbor of ours, and the next planet out, Jupiter. This planet never gets much closer to us than about 500 million miles (800 million km). The enormity of this distance detered even NASA from attempting exploration by probe until 1972. Then it despatched Pioneer 10 to Jupiter, the first of two nearly identical craft; its twin, Pioneer 11, was launched a year later.

The design of these craft was radically different from that of earlier probes. The most prominent feature was a dish antenna 9 feet (2.7 meters) across, needed to relay directional signals back to

Left: The Pioneer 10 and 11 probes pioneer investigation of the giant planet Jupiter in 1973 and 1974. Their imaging systems send back the best detail yet of Jupiter's banded atmosphere, as well as its most prominent feature, the Red Spot.

Right: When the Voyager probes encounter Jupiter in 1979, they send back superlative pictures of the planet, including this one of a nearly 'full' Jupiter.

Below: Seen in close-up and pictured in false color to show up details is Jupiter's Red Spot. Measuring up to 17,000 miles (28,000 km) long and 9000 miles (14,000 km) wide, it appears to be a permanent Jovian super-hurricane.

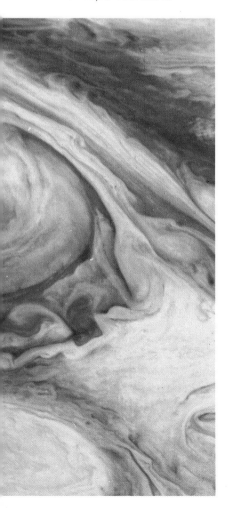

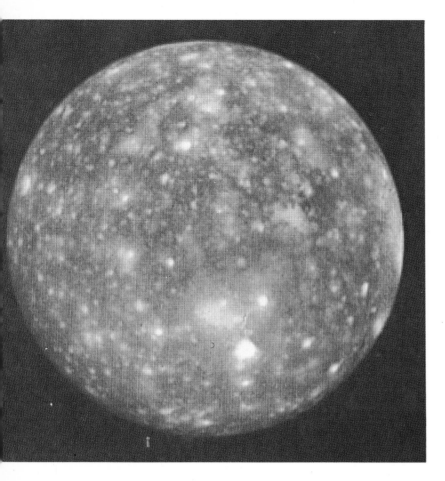

Earth. The power source was also quite novel. Solar cells could not be used because in deep space the intensity of sunlight is too low to generate enough electricity. So Pioneer used nuclear power from units called radioisotope thermoelectric generators (RTGs). They converted the heat from the radioactive decay of plutonium into electricity by means of thermocouples.

Another potential 'barrier' to space probes journeying to the outer planets is the asteroid belt, a broad band of space in which orbit thousands upon thousands of mini-planets, boulders and rocks of all shapes and sizes. Both Pioneers, however, survived passage through the belt and made their encounters with Jupiter in December 1973 and 1974, respectively; after traveling some 600 million miles (1000 million km). They took the first close-up pictures of this giant among planets, which is big enough to contain 1300 Earths. They found no evidence of any solid surface under the multicolored cloud belts that ringed the planet — Jupiter seemed to be a great ball of gaseous and liquid hydrogen. They did, however, detect a magnetic field, vastly more powerful than the Earth's.

Incredible images

The two Pioneers approached Jupiter with quite different trajectories, and when they swung behind the planet they were flung by the gravity sling-shot effect in different directions. Pioneer 10 headed out of the solar system, while Pioneer 11 headed towards the next planet out, Saturn. It reached this glorious ringed planet in 1979 after a five-year journey, discovering an additional ring and another satellite to those already known.

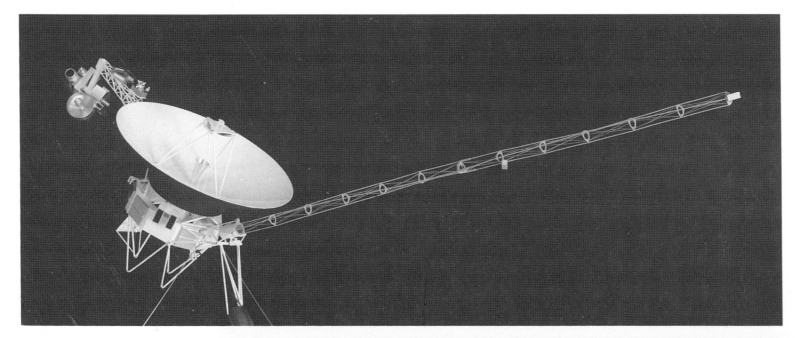

Left: The Voyager probes are of identical design. They have a 12 foot (3.7 meter) diameter dish antenna for transmitting and receiving signals to and from Earth. The cameras and radiation and particle detectors are clustered together on a boom at the top of the picture. Mounted on a strut beneath the dish is the nuclear generator that produces electricity. The long gold-colored boom carries magnetometers.

Far left: Callisto, one of Jupiter's 'big four' moons. Unlike the other three, it has a very ancient cratered surface. With a diameter of some 3000 miles (4800 km), it is similar in size to the planet Mercury.

Left: Quite unlike Callisto and indeed any other body in the solar system is the somewhat smaller moon Io. It is a brilliant yellow-orange in color, due to a covering of sulfur. This material is spewed out by active volcanoes. Voyager 1 spotted a volcano errupting on Io's limb, sending sulfur and dust particles to a height of over 125 miles (200 km).

Right: This photo montage from Voyager pictures shows the glorious ringed planet Saturn and the six largest of its 20 plus moons. In the foreground is Dione; Tethys is next right, and then comes Mimas. At top right is Io. Above Dione in order are Enceladus and Rhea.

The Pioneer missions blazed a trail into the outer solar system for two much more advanced robot explorers called Voyager, which were launched in 1977. Like Pioneer 11, both Voyagers, 1 and 2, visited Jupiter (both in 1979) and Saturn (1 in 1980, 2 in 1981) in turn. Equipped with superior imaging equipment, they sent back the most amazing pictures of the Space Age. At Jupiter they showed incredible views of the swirling atmosphere in kaleidoscopic colors and of the enigmatic Red Spot, the eye of a furious storm that has been raging for centuries, maybe even millennia. At Saturn the probes spied wind belts coursing through the atmosphere at speeds up to 1100 mph (1800 km/h).

The Voyager pictures of the moons of the two giant planets were equally revealing. They showed that the four biggest 'Galilean' moons of Jupiter are all quite different in appearance. The most startling is Io, which must rank as the most colorful moon in the whole solar system. On its bright orange surface live volcanoes spew out molten rivers of sulfur. The Voyagers spotted no fewer than eight volcanoes in eruption. They also discovered that both Jupiter and Saturn have tiny moons that we cannot see from Earth. They spied at least three new moons of Jupiter and no less than 11 of Saturn!

Some of Saturn's new moons have been called shepherd moons because they appear to be keeping the particles in Saturn's rings in place. And it was Saturn's rings in the end that stole the Voyager's show.

From Earth we can see three broad rings that girdle the planet's equator. The Voyager pictures revealed that these rings are in fact made up of hundreds upon hundreds of individual ringlets. They also discovered several more ring systems both inside and outside the rings that are visible from Earth.

Left: Saturn and two of its moons pictured here from 8 million miles (13 million km). At the top is Tethys, which is casting a shadow on the planet's surface at lower right. The other moon is Dione. Notice the gap in the two main rings. It is known as the Cassini division after the Italian who discovered it.

Right: The Voyager pictures show startling detail of the system of rings girdling Saturn's equator. They prove to consist of thousands of separate ringlets, composed of orbiting particles of ice and rock.

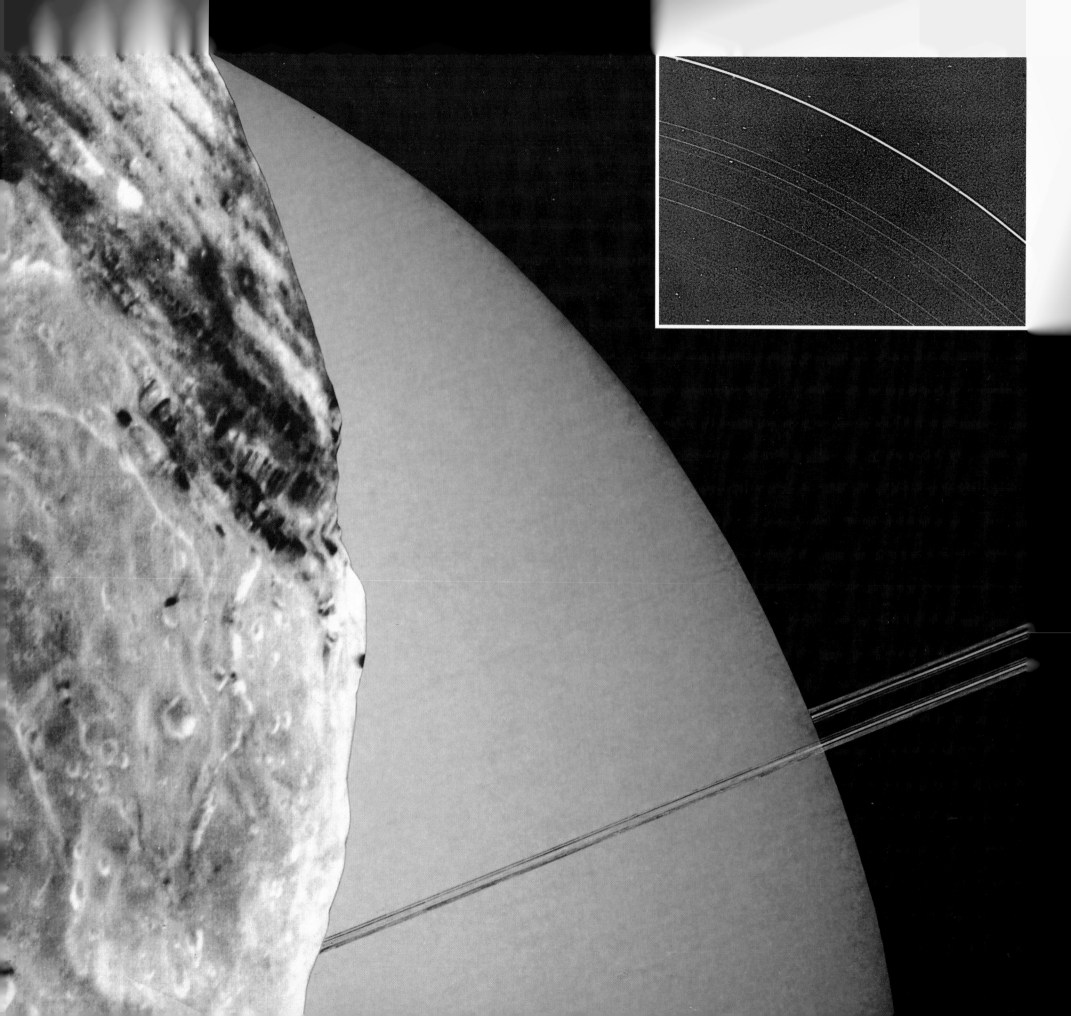

Left: Two images taken by Voyager 2 at Uranus in January 1986 are superimposed in this picture. At left is the rugged icy surface of the 500-km (300-mile) diameter moon Miranda. Beyond is the blue-green expanse of Uranus itself, some 105,000 km (65,000 miles) away. The rings that girdle the planet have been accentuated artificially.

Inset left: The faint ring system around the equator of Uranus is revealed in this Voyager 2 picture. By far the brightest ring, the epsilon ring, is the outermost ring in the system.

Right: This is the Uranian moon Umbriel, some 1200 km (750 miles) in diameter, which orbits about 270,000 km (170,000 miles) from the planet. Its most prominent feature is a curious white ring, about 140 km (90 miles) in diameter. It could be a circle of frost around a recent impact crater.

Journey to Uranus

The Voyager probes left Saturn on different trajectories. Voyager 1 began heading out of the solar system, but Voyager 2 had other calls to make before its mission was over. It swung round Saturn in the late summer of 1981 and began heading for a distant rendezvous with the next planet out, Uranus, timed for January 1986. It arrived exactly on schedule and in near-perfect working order for a close encounter on the 24th.

On that day Voyager 2 swooped within about 50,000 miles (80,000 km) of the planet's cloud tops. Streams of new data flowed from it to Voyager mission control at the Jet Propulsion Laboratory (JPL) in California, adding to the wealth of data that had already been received over the previous few days. Voyager's magnetometer detected a weak magnetic field, about one-third as strong as the Earth's. The Voyager images showed clouds coursing through the greenish-blue atmosphere of the great gaseous planet. They revealed a tenth ring around the planet's equator, and ten new tiny moons, adding to the five large moons we can spot from Earth.

Close-up pictures of these large moons show remarkable features. There are high mountains on Oberon, and its surface is also marked with icy craters, as is Titania's. But the most remarkable moon is Miranda, pictures of which, said JPL, 'are really mind-blowing'. Miranda has geological features quite unlike those of any other body in the solar system. There are deep canyons, mysterious scars and chevron features, streaked expanses of black material and puzzling angular features.

As Voyager whipped around Uranus, the planet's gravity changed its direction and put it on course for yet another planetary rendezvous three and a half years hence. The target is Neptune, at present the outermost planet in the solar system. The estimated time of arrival at Neptune is 24 August 1989.

Heading for the stars

Pioneer 10's encounter with Jupiter occured late in 1983, but its usefulness was not yet over. It continued to monitor the conditions in interplanetary space as it journeyed farther and farther out. In June 1983 it crossed the orbit of Neptune, then the outermost planet, and left the solar system. It began heading out into interstellar space, perhaps, who knows, to rendezvous with another planet of another solar system some time in the distant future.

The chances that Pioneer will encounter another planet and that that planet will harbor intelligent life are infinitesimally remote, but it could happen. So to inform those aliens that might find it, where it came from and who sent it, Pioneer carries a pictographic plaque. Diagrams, sketches and binary numbers provide the information. A naked man and woman, drawn to scale with the spacecraft, show the dominant life form on the planet from whence it came.

Pioneer 11 and both Voyagers will also eventually disappear into interstellar space. The former has a plaque like its sister craft, but the Voyagers do better than this and carry a record disc. Called 'Sounds of Earth', the disc holds greetings from peoples of the Earth in 60 languages, and evocative sounds of the natural and man-made world. It also carries electronically coded photographs showing the multifaceted nature of our planet. Provided the far-off aliens that find it can fathom out how to play it (helpful instructions are depicted on the record cover), what a treat they have in store!

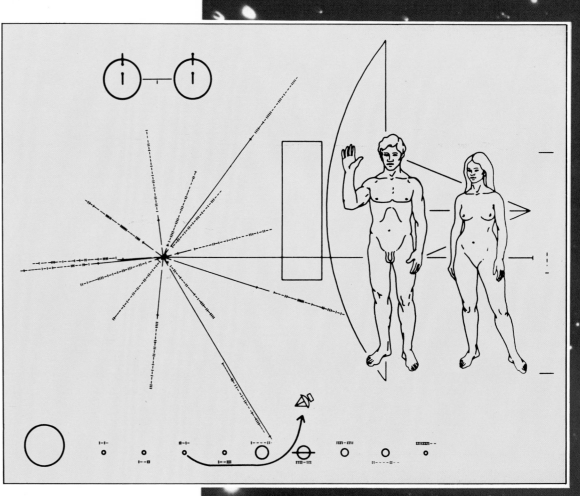

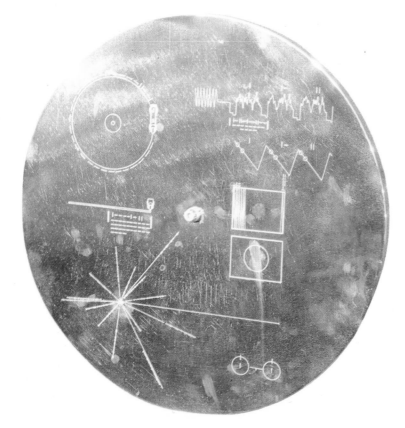

Above: Just in case the Pioneer probes do eventually come upon intelligent life in our galaxy, they carry a plaque to identify who sent them.

Left: The Voyager probes carry a record disc housed beneath a cover like this. The diagrams etched on the cover give coded instructions about how to play the disc.

Right: No-one yet knows for certain whether there are any other living, reproducing beings 'out there' in the depths of space. Statistically it is almost certain that there are. Out of the billions upon billions of stars in our own galaxy alone there must be many that have planets which enjoy a climate similar to our own. Why shouldn't life have evolved on these planets too?

Chapter 8

Welcome to NASA

One thing that has characterized the American space program from the beginning has been the 'open-house' policy of NASA. The fullest possible information has always been available to the press and the public alike concerning virtually every aspect of the space program – good and bad. Only when a launch is classified, as most military launches are, does the NASA public relations machinery grind to a halt.

As part of this open-house policy many NASA centers positively encourage visitors. There are, throughout the United States, seventeen NASA establishments, and nine of these have visitors centers open to the public on most days. It is no surprise that the Kennedy Space Center tops the popularity poll among visitors, attracting several million every year, as befits the world's premier spaceport. The facilities at the visitors center there are second to none, the exhibits and non-stop film presentations are enthralling, and the tours around the launch sites are not to be missed. There is also of course always the chance of the added bonus of a shuttle launch or landing, both of which are unforgettable experiences.

The other two main operational centers for space flights are the Johnson and Goddard Space Centers. Johnson is the major training facility for astronauts and is the home of manned space flight mission control. Goddard is among other things the nerve center for NASA worldwide communications network.

Kennedy, Johnson and Goddard are devoted almost exclusively to space operations, as is the historic Marshall Space Flight Center. But several other centers are concerned as much with flight through the atmosphere as through space. They had their origins in NACA, the National Advisory Committee for Aeronautics, the body that preceded NASA. The main centers for NASA's aeronautical research are the Langley Research Center, the Ames Research Center, the Dryden Flight Research Facility and the Lewis Research Center.

Some details of the work carried out at the main NASA centers appears below. Their locations are shown on the accompanying map on page 218.

The administrative center for NASA is the Headquarters in Washington DC. Dr James Fletcher is currently NASA Administrator. He and some 1300 staff administer the NASA budget, which for the late-1980s was some $8 billion per year.

NASA HQ exercises management over the aerospace centers throughout the country and is responsible for the overall planning

These pages: A now familiar scene at the world's premier spaceport, the Kennedy Space Center in Florida. The spectacular aerial view of a shuttle take-off shows to good effect the typical swampy alligator-infested landscape.

of projects. The projects are organized within six program offices. The Office of Aeronautics and Space Technology is responsible for aerospace research and technology programs. The Office of Space Flight is responsible for manned space flights, which involves mainly shuttle and Spacelab flights. The Office of Space Science and Applications is concerned with satellites and deep space probes. The Office of Space Station is directing the oncoming space station program. The Office of Space Tracking and Data Systems deals with tracking and communications and the acquisition and dissemination of data. The Office of Commercial Programs encourages the commercial application of space technologies.

Ames Research Center/Dryden Flight Research Facility

Ames and Dryden research centers have been amalgamated since 1981. The installation that was formerly Ames is now referred to as Ames-Moffett. It is located at the southern end of San Francisco Bay, near the US Naval Air Station at Moffett Field.

In the field of space exploration Ames has responsibility for the Pioneer series of probes, one of which, Pioneer 10, became the first spacecraft to leave the solar system in 1983. It is more involved, however, in aeronautical research. It has a variety of wind tunnels, providing a speed range from subsonic — below the speed of sound, to hypersonic — more than five times the speed of sound.

Among the projects Ames is involved in is the quiet short-haul jet, a design for a transport operating from short runways close to city centers. It employs over-the-wing engines and an innovative wing and flap arrangement to create very high lift. Ames also conducts research with rotary wing aircraft. This includes work on new types of rotors for helicopters, as well as hybrid craft such as the tilt-rotor aircraft. This has fixed wings with engines at the ends that can be tilted vertically to turn the propellers into helicopter-like rotors for vertical take off and landing.

A much more conventional aircraft forms the Kuiper Airborne Observatory, which is based at Ames-Moffett. It carries a 36 inch (91 cm) telescope that records images in the infrared region of the spectrum. Flying at an altitude of 39,000 feet (12,000 meters), it is above the water vapor in the atmosphere that absorbs infrared radiations and prevents observations at these wavelengths from the ground.

Ames-Dryden is located at Edwards Air Force Base in the Mojave Desert, some 60 miles (100 km) north of Los Angeles. It is an ideal location for flight testing, being remote and enjoying excellent weather all year round. In aviation it is noted for its work in out-of-the ordinary advanced airplane design in such projects as the oblique-wing, HiMAT and the forward-swept wing.

The oblique-wing craft has a wing that extends conventionally at right-angles to the fuselage for take off, but then pivots around the center during forward flight, with one half of the wing swept forward and the other half swept back. In theory this should considerably reduce drag at high speeds.

HiMAT (highly maneuverable aircraft technology) was a

Right: This aerial shot of the Gerard Kuiper Airborne Observatory, based at Ames, shows the opening in the fuselage through which the infrared telescope is pointed. With a mirror diameter of 36 inches (91 cm), it is the largest airborne telescope in the world.

Below: A model of a V/STOL (vertical/short take-off and landing) fighter being tested in Ames-Moffett's transonic wind tunnel. Five V/STOL concepts are being evaluated to build up a comprehensive aerodynamic data base for this type of aircraft.

Below: Another fascinating Ames-Moffett project centers on the rotor systems research aircraft (RSRA), illustrated here. It can be configured to fly as a pure helicopter, a fixed-wing aircraft or, as here, a compound helicopter with wings as well as a rotor.

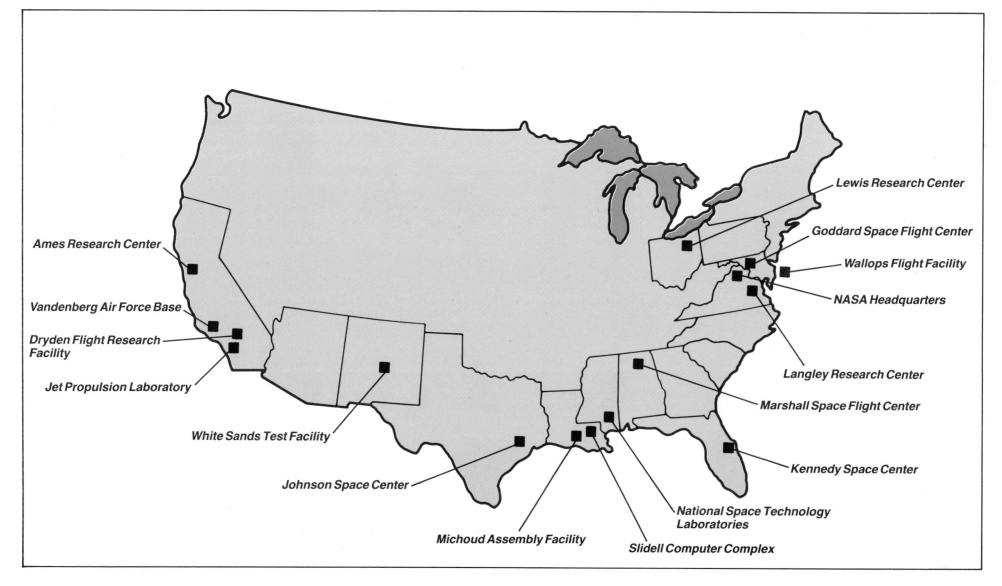

Ames Research Center

Vandenberg Air Force Base

Dryden Flight Research Facility

Jet Propulsion Laboratory

White Sands Test Facility

Johnson Space Center

Michoud Assembly Facility

Slidell Computer Complex

National Space Technology Laboratories

Kennedy Space Center

Marshall Space Flight Center

Langley Research Center

NASA Headquarters

Wallops Flight Facility

Goddard Space Flight Center

Lewis Research Center

remote-control flying testbed for future fighter design. It featured a secondary or canard wing in front of the main wing, which also had vertical winglets at the tips. The maneuverability of the design proved breathtaking. The forward-swept wing design, which has its wings projecting forwards, also aims for greater maneuverability. Ames-Dryden is currently testing the prototype aircraft, the X-29A.

On the space side, Ames-Dryden is responsible for shuttle landings at Edwards Air Force Base, which is an alternative landing site to Kennedy. Edwards is generally used for Spacelab flights because of the extra landing load, since there is plenty of room there for the shuttle to overshoot if it needs to.

Goddard Space Flight Center

The Robert H. Goddard Space Flight Center was the first new facility opened by NASA in the year following its formation (1958). It is now one of the biggest NASA establishments, with a workforce of more than 8000 people. The management of Wallops Flight Facility and the Space Telescope Science Institute is also the responsibility of Goddard.

Goddard is located at Greenbelt, Maryland, just 10 miles (16 km) north-east of the nation's capital. Its prime function is as the nerve center for NASA's worldwide space communications network, NASCOM. It receives satellite telemetry and communications via a worldwide network of tracking stations known as the Satellite Tracking and Data Network (STDN). Stations within this network have been gradually phased out as the geostationary tracking and data relay satellites (TDRSs) have become operational. The TDRSs pick up data from spacecraft, including the space shuttle, and beam it down to White Sands Facility in New Mexico, which channels the data to Goddard.

Goddard is the prime receiving and processing center for US satellite data. It receives data continuously from weather and Landsat satellites. The data is stored in a massive computer complex, forming the National Space Science Data Center.

Jet Propulsion Laboratory

The Jet Propulsion Laboratory (JPL) is located at Pasadena, California, some 20 miles (30 km) to the north-east of Los Angeles. It is operated for NASA by the California Institute of Technology

Above right: The operations control room of the Goddard Space Flight Center at Greenbelt, Maryland. Goddard is the nerve center for NASA's satellite communications and tracking network worldwide.

Right: With its 210 foot (64 meter) diameter dish pointing up to the heavens, the Goldstone antenna of JPL's Deep Space Network tunes in to the faint signals emitted by probes billions of miles away.

Above: The locations of NASA establishment in the United States.

(Caltech) and occupies a special place in NASA history because it built the nation's first satellite, Explorer 1. Its main responsibilities these days, however, are for deep-space, not near-space missions.

JPL designs, tests and operates deep-space probes. It designed and built the incredibly successful Voyager probes, which have revolutionized study of the outer planets, most recently Uranus (1986). It has also built the Galileo probe that will orbit Jupiter and release a capsule into its atmosphere.

As part of its deep space program JPL developed, and now operates, the Deep Space Network, which 'listens' with huge dish antennas for the faint signals coming back from space probes maybe billions of miles away. The DSN uses tracking stations at Goldstone in California, at Madrid in Spain, and at Canberra in Australia. The nerve center for deep space operations is JPL's mission control room, which comes alive when probes move in for planetary encounters.

Johnson Space Center

The Lyndon B. Johnson Space Center was established in 1961 as NASA's primary manned space flight center, being responsible for the design, development and testing of spacecraft; the selection and training of astronauts and the planning and conducting of manned space missions. It is located some 20 miles (30 km) south-east of downtown Houston in Texas.

The best-known role of Johnson to the world at large is as mission control for American manned space flights, with the famous call sign, 'Houston'. Mission control takes over control of every shuttle flight as soon as it has lifted off the pad at the Kennedy Space Center. During every mission, shifts of flight controllers at mission control operate rows of communications consoles round the clock. Data and TV images from the orbiting shuttle are projected on screens before them. A running plot on a world map shows the location of the craft over the Earth.

Among the training facilities at Johnson that prepare astronauts for flights is the space shuttle mission simulator. This is similar to, but more complicated than, the simulators pilots use for airplane flight training. Another major training aid is the water tank, or neutral buoyancy simulator, in which astronauts practice EVA procedures in weighted spacesuits. This gives a good approximation to the weightless conditions the astronauts will experience in orbit.

With an eye to the future, Johnson is the lead center for the development of the NASA space station for the 1990s. It has specific design responsibilities for the interfaces between the space station and the space shuttle, which will service it.

Johnson also manages NASA's White Sands Test Facility at Las Cruces in New Mexico. The Facility is located on the western slopes of the San Andres Mountains at the edge of the US Army's White Sands Missile Range. It houses laboratories and test stands for testing elements of the shuttle's propulsion and power systems. The ground station for receiving communications from the tracking and data relay satellites (TDRSs) is also located at White Sands.

Kennedy Space Center

The John F. Kennedy Space Center is located just inland from Cape Canaveral on Merrit Island, some 50 miles (80 km) east of the city of Orlando. Since the dawn of the shuttle era, it has become busier than ever before.

The Kennedy Space Center is the main launch site for the space shuttle, and a landing site too. Another launch and landing site has been built at the Vandenberg Air Force Base in California, but is currently not in use. At the Kennedy Center shuttle operations are concentrated on the sprawling Complex 39. This launch complex was built to support the Apollo Moon-landing missions in the 1960s and 1970s. It underwent modifications to fit it for its new role with the shuttle.

Dominating Complex 39 is the Vehicle Assembly Building (VAB), one of the largest buildings in the world. Over 525 feet (160 meters) tall, it covers an area of 8 acres (3 hectares). It is in this great edifice that the shuttle stack of orbiter, external tank and solid

Below: An unused Saturn V Moon rocket, 365 feet (111 meters) in length, dominates the 'rocket park' at the Johnson Space Center at Houston, Texas. Note it is split up into its main sections – the Apollo spacecraft and three rocket stages.

Right: A young visitor to Johnson's comprehensive Visitors Center gazes in awe at the command module used on the last Moon-landing mission, Apollo 17 (1972). Note the blackened base.

Far right: Visitors to the Kennedy Space Center get their first glimpse of rocketry just before they cross over to Merritt Island.

JOHN F. KENNEDY SPACE CENTER
MERCURY-REDSTONE
FIRST U.S. MANNED SPACE VEHICLE
SUBORBITAL MISSIONS
ALAN SHEPARD   MAY 5, 1961
VIRGIL GRISSOM   JULY 21, 1961
SPECIFICATIONS
HEIGHT: 83 FT. THRUST: 78,000 LBS. WEIGHT: 33 TONS
PROPELLANTS: ALCOHOL & LIQUID OXYGEN

rocket boosters is put together. The orbiter arrives in the VAB via the orbiter processing facility (OPF), two low buildings to the west of the VAB. In the OPF the orbiter is checked out after a mission, serviced and prepared for the next. New payloads may also be installed there.

To the east of the VAB is the four-story building of the launch control center (LCC). This is in charge of the checkout of shuttle systems during the countdown and launch operations. Immediately the shuttle has left the pad, control reverts to mission control at Houston. The LCC looks out to the launch pads along the turnpike-wide crawlerway down which the shuttle stack is transported. Pad A, the first one to become operational, lies some 3.5 miles (5.5 km) away. Pad B lies about 1.5 miles (2 km) north of Pad A. The shuttle stack is serviced on the pad from a fixed tower

and a rotating structure that provides access to the orbiter for the installation and servicing of payloads at the pad.

The shuttle is transported on its launch platform from the VAB to the pad by a gigantic eight-crawler tracked vehicle. The shuttle remains bolted to the launch platform until lift-off. Just before and during lift-off the platform is flooded with water at a peak rate of 900,000 gallons (3.5 million liters) a minute to reduce the sound reflections that could damage the orbiter and its payload.

Other facilities at the Kennedy Center include a computer complex; a satellite communications switching center that links Kennedy with Houston and other NASA centers; a headquarters building, which is the administrative center; and the operations and checkout building, where Spacelab and other shuttle payloads are checked out prior to installation.

Above: The main 'rocket park' at the Kennedy Space Center, containing a collection of early launching hardware.

Right: Another popular Kennedy exhibit, a full-size replica of the Apollo lunar landing module. Photography at the Center is positively encouraged.

Regular guided bus tours take visitors around the Center throughout the day. They may also go across the Banana River to the Cape Canaveral Air Force Station, where military and unmanned NASA launches still take place. Of historic interest there is the Air Force Space Museum, which contains a fascinating variety of early space hardware.

Another interesting aspect of the Center is that it is located in the midst of a wildlife haven – the Merritt Island Wildlife Refuge. The tight security surrounding the Space Center also has the added benefit of protecting the habitats of some 285 species of birds, 25 species of mammals, 117 species of fish and 65 different types of amphibians and reptiles. Species include the American bald eagle, the nation's emblem, ibis, heron, armadillo, turtle, manatee and of course alligators. Alligators can almost always be seen in the channels and lagoons that criss-cross the marshy expanse that makes up Merritt Island.

Langley Research Center

This is NASA's oldest research center, which has a long history of aerospace innovation. It is located at Hampton, Virginia, some 100 miles (160 km) south of Washington DC. Its main work these days centers on aeronautical research, especially on advanced concepts for future aircraft. Its facilities include numerous wind tunnels and a structure for deliberately crashing full-size aircraft.

In the space field, it was responsible for the Viking Mars program in the 1970s. Currently, it manages the Scout launching vehicle, often fired from the nearby Wallops Flight Facility. One of its recent projects was the Long Duration Exposure Facility, lofted into space by the shuttle in spring 1984 with more than 50 experiments on board. Its recapture and return to Earth was delayed as a result of the grounding of the shuttle fleet following the *Challenger* disaster in 1986.

Lewis Research Center

Lewis is one of the most versatile of NASA's field stations, located in Cleveland, Ohio on the west of Cleveland Hopkins Airport. Established in 1941 as a NACA research center, it came under NASA jurisdiction in 1958. Lewis conducts research in aircraft propulsion, space propulsion and power systems, and energy systems.

One of Lewis's prime programs in aircraft propulsion is concerned with advanced turboprop design. This new type of turboprop engine will be able to power aircraft to the same speed as jets but with much better fuel economy. The main feature of the design is a multiblade swept-tip propeller. In space propulsion Lewis is experimenting with ion thrusters for long-duration, low-thrust motors for deep space missions. It also has responsibility for NASA's most powerful expendable space launch vehicles, the Atlas-Centaur and Titan-Centaur.

Lewis is also the nation's leading center for terrestrial alternative energy research involving solar energy, wind energy, and fuel cells. Solar energy power schemes developed by Lewis are at work throughout the world, mainly in developing countries. In Tunisia, for example, solar cell arrays provide the total electrical power for the village of Hamman Bradha, some 90 miles (145 km) from the nation's capital, Tunis.

Lewis's wind-energy programs are in evidence much nearer to home. Massive wind turbines of characteristic propeller-type design are operating at their large-scale test facility at Plum Brook Station near Sandusky, Ohio, and at several other locations throughout the country. Near Medicine Bow, Wyoming, for example, is one of the most powerful wind turbines yet developed. With a 256 foot (85 meter) long rotor blade, it produces a peak 4000 kilowatts of electricity. It could lead to the construction there of a 'wind farm' with as many as 40 turbines.

Marshall Space Flight Center

The George C. Marshall Space Flight Center is located within the US Army's Redstone Arsenal at Huntsville, Alabama. It was the nation's pioneering space center for the development of early space launch vehicles under the direction of Wernher von Braun. They included the Saturn series of rockets, the biggest of which, Saturn V, was instrumental in sending Apollo spacecraft to the Moon and Skylab into orbit.

Currently, Marshall has responsibilities for the propulsion hardware for the space shuttle – solid rocket boosters, main engines and external fuel tank – as well as for major shuttle payloads. These include the European-built Spacelab and the Hubble space telescope. Spacelab made its debut in 1983, being seen perhaps as a prototype laboratory module for the 1990s space station, which Marshall is currently designing. The space telescope, which is set to revolutionize optical astronomy, was ready for launch by 1986.

Marshall is also a training center for astronauts and engineers working on space flight hardware. One of its most impressive training facilities is the giant water tank, or neutral buoyancy simulator, in which astronauts can simulate weightlessness. It is so telescope seems certain to revolutionize astronomy when it becomes fully operational.

Marshall also manages two other NASA outstations, the Slidell Computer Complex and the Michoud Assembly Facility, both near New Orleans, Louisiana. Michoud is mainly concerned with the manufacture of the shuttle's external tank.

Close-by the Marshall Space Flight Center in Huntsville is the Alabama Space and Rocket Center (ASRC), which proudly boasts that it is the 'Earth's largest space museum'. Owned and operated by the State of Alabama as a non-profit educational institution, it acts as the official visitors center for Marshall. The ASRC has a comprehensive collection of exhibits covering all phases of the US space program past and present and a marvellous rocket park, including one of only three Saturn V rockets still in existence (the other two are at the Johnson and Kennedy space centers). Until her death in 1984, pride of place among the inside exhibits went to 'Miss Baker', the squirrel monkey that pioneered primate spaceflight back in May 1959.

Vandenberg Air Force Base

NASA's prime launch sites are at the Kennedy Space Center for manned missions and Cape Canaveral for unmanned flights. It also has launch facilities on the other side of the continent in California at Vandenberg Air Force Base, located on the coast about 150 miles (250 km) north-west of Los Angeles. The NASA facilities at

Vandenberg are often termed the Western Test Range (as opposed to Cape Canaveral, which is called the Eastern Test Range).

Scores of launches have been made from the Vandenberg site since it first became operational in 1960. These have been of satellites that need to go into a polar orbit; one that takes them over the North and South Poles. This kind of orbit is required for many weather satellites and Earth-resources satellites such as Landsat, which must scan the whole of the Earth's surface repeatedly. These payloads have been lifted into orbit by launch rockets such as the Scout and Delta.

A shuttle launch and landing site has also been built at Vandenberg for launching military and civilian payloads that require a polar orbit. It was scheduled to become operational in 1986, but was mothballed after the *challenger* disaster.

The facilities for orbiter processing and launch preparation at Vandenberg are quite different from those at the Kennedy Space Center. But the two facilities do have something in common – they were modifications of existing structures. Those at Complex 39 at Kennedy were based on the Saturn V/Apollo facilities. Those at

Below: This monument marks the site from which the first American astronauts were launched into orbit in the 1960s. The '7' signifies the original seven Mercury astronauts.

Right: Shuttle hardware on the move to the launch pad. A wide-angle lens is needed to squeeze the shuttle stack and the towering VAB into the same frame.

Vandenberg were based upon the nearly completed facilities built for the Air Force's Manned Orbiting Laboratory (MOL) program, which was canceled in 1969. They form space launch complex 6 (SL-6), colloquially known as 'Slick Six'.

The launch site at Vandenberg could not differ more from that at Kennedy. It is hemmed in on three sides by the Santa Ynez mountains, contrasting with Kennedy's pancake-flat marshy terrain. The largest building in the launch complex is the payload preparation center, which is fixed, unlike the three other major structures, the mobile service tower (MST), the shuttle assembly building (SAB) and the payload changeout room (PCR), which can move back and forth along railroad-type tracks.

Assembly of the shuttle stack takes place at the launch pad on a launch mount. First the MST moves up to the pad and lifts the two solid rocket boosters into position. Next the SAB moves in from the other side of the pad and mates with the MST to form an enclosed weatherproof space, inside which assembly and payload installation are completed. The payload is transferred to the pad via the PCR, which travels from the payload preparation center into the shuttle assembly building. Immediately prior to launch the MST and SAB are rolled back and the shuttle is fueled and crewed from a gantry-type access tower.

Lift-off occurs in a southerly direction, so that the boosters can be jettisoned over the ocean south-west of Los Angeles. The remains of the external tank, broken up by re-entry, also impact in the Pacific Ocean or over unpopulated Antarctica. The shuttle orbiter comes in to land some 16 miles (25 km) north of the launch

Left: Among the impressive display of space hardware assembled at the Alabama Space and Rocket Center is this six-wheeled prototype for the Apollo lunar buggy. In the background is an Atlas rocket.

Above: At the Marshall Space Flight Center astronauts train for weightlessness inside a huge water tank, the neutral buoyancy chamber. Their space/diving suits are weighted so that they achieve neutral buoyancy, a condition in which they neither float nor sink.

Above: Wallops Flight Facility is the main launch base for the Scout solid-propellant rocket. Here the rocket is being raised into its vertical position prior to launching. Note the flag on the nose cone, which signifies that it is carrying a British satellite, UK-6.

site on an existing runway that had to be nearly doubled in length to 15,000 feet (5000 meters).

Wallops Flight Facility
Managed by Goddard, the Wallops Flight Facility was established in 1945 as a NACA research center attached to Langley. It is located on the Atlantic coast of Virginia some 40 miles (65 km) south-east of Salisbury, Maryland.

Wallops is one of the world's oldest rocket launch sites. Some 100-150 launches take place there each year. Scout rockets are used to launch small satellites into Earth orbit, as well as sounding rockets which carry instruments into the upper atmosphere and near space.

Wallops manages the National Sounding Rocket Program and also gives assistance to foreign countries wishing to set up their own sounding rocket facilities. In addition, it coordinates NASA's scientific balloon projects, which are centered on a launch facility at Palestine in Texas.

Chapter 9

The Way Ahead

On Tuesday 28 January 1986 the unbelievable happened. Shuttle orbiter *Challenger* exploded in the skies above Cape Canaveral. Commented *Time* magazine: 'In a fiery instant the Nation's complacent attitude to space flight evaporated'. NASA, after nearly three decades of unbridled excellence and seemingly divine infallibility, had been proved fallible after all.

Privately and publicly, but above all deeply, America mourned the death of its seven astronaut martyrs who, said President Reagan, 'had a hunger to explore the universe and discover its truths'.

Messages flooded in from around the globe. In the Pope the tragedy 'provoked deep sorrow in my soul.' Britain's prime minister Margaret Thatcher observed that: 'New knowledge sometimes demands sacrifices of the bravest and the best.' Commented Toronto's *Globe and Mail* on this theme: 'Glory and adventure often go hand in hand with danger and death.' Soviet cosmonauts, more aware than any of the hazards of their profession, sent messages of sympathy. Russia named two craters on Venus after Christa McAuliffe and Judy Resnik.

Then, the mourning was over. It was time to pick up the pieces. The President appointed a 13-member commission to investigate the causes of the accident and the circumstances that may have led up to it, and to make recommendations to prevent anything similar ever happening again. The members of the commission included engineers, scientists and astronauts Neil Armstrong and Sally Ride. It was chaired by former Secretary of State William Rogers. The commission were given just four months to report back to the President.

Flawed decision-making

Even before the Rogers commission got properly to work, the immediate cause of the *Challenger* disaster was becoming evident. It was the failure of the O-ring seals in the bottom joint of the casing of the right-hand SRB (solid rocket booster). These synthetic-rubber seals are designed to prevent the searing hot gas generated inside the rocket from escaping through the joint. But on *Challenger*'s last lift-off, that is precisely what happened, triggering off the catastrophe.

As the commission's hearings progressed, an extraordinary story began to emerge about the O-rings, about the lack of

A scene in orbit at the turn of the century. A shuttle orbiter maneuvers in to dock with the space station, now a thriving research and servicing facility.

In the background is one of the free-flying platforms carrying a variety of instruments for probing the space environment.

communication between engineers and decision-makers, and about 'flawed decision-making' at many levels. The checkered history of the O-ring seals, the commission decided, also epitomized a malaise within the Agency itself.

As early as 1979, two years before the shuttle first took to the skies, the O-rings were seen as a problem. In 1980 a NASA engineering panel called them 'inadequate'. Three years later they were put on a 'criticality 1' list, meaning that they lacked redundancy (back-up) and their failure would lead to 'loss of mission and crew'.

After SRBs from several early shuttle flights showed signs of O-ring erosion, NASA asked the booster contractor Morton Thiokol to seek a solution. Thiokol set up a seals task force. After shuttle 51-C launched in January 1985, following overnight temperatures in the 20s (F), O-ring damage was found to be extensive. The reason was plain, at low temperatures the synthetic rubber loses its resilience and cannot properly expand into the joint. Thiokol's engineers warned of the vulnerability of the seals, of possible catastrophe. Some, it was said, 'held their breath' during every shuttle launch in case of O-ring failure.

The temperature on the night before *Challenger*'s fateful launch also plummeted to the 20s. Thiokol engineers opposed launching following such lows. Their advice was ignored by higher management. With ice coating the launch gantry, shuttle builder's Rockwell opposed the launch because of the danger of ice impact on the orbiter. Their advice was ignored. No-one communicated reservations about launching to NASA's top decision-makers, such as Jesse Moore. As a result, the launch was allowed to go ahead, and *Challenger* blasted off – into oblivion.

Rogers recommends

Early in June the Rogers commission presented a 256-page report to President Reagan, summarizing nearly 3000 pages of transcripts taken during interviews with 160 people. As expected, the report criticized NASA management structure and the lack of communication within it. The report confirmed that the specific cause of the *Challenger* disaster was failure of the O-ring seals and recommended that a new design be produced.

Among the other things it recommended were study of a crew escape system and improvements in tires, brakes and steering. There should always be sufficient spares available, the report said, to prevent one shuttle being cannibalized so that another can fly, as had often happened in the past. Edwards Air Force Base in California should become the prime shuttle landing site rather than Kennedy, where the weather is notoriously unpredictable.

President Reagan then placed the ball firmly in NASA's court and charged newly appointed NASA Administrator Dr James Fletcher to implement the commission's recommendations. The President's appointment of Dr Fletcher for a second stint as NASA Administrator was itself a step in the right direction. The first time round, between 1971 and 1977, he had lifted the Agency out of the doldrums of the post-Apollo era and thrust it vigorously towards the shuttle age.

Early in 1987 in an upbeat speech in Los Angeles, Dr Fletcher affirmed that NASA and the American space program were back on course. 'NASA has rededicated itself,' he said, 'to giving the kind of space program Americans deserved and can be proud of. We are totally dedicated to returning the nation to safe and reliable space flight.' The management of the Agency had been restructured,

Bottom left: Ice festoons the gantry on launch pad 39B on the eve of Challenger's *fateful launch. Warnings not to launch because of the extreme cold are ignored.*

Below: Just seconds after lift-off, a tell-tale puff of smoke emerges from the lower joint in the right-hand SRB, indicating that the O-ring seal is not sealing.

Right: An SRB streaks away from the fireball that has just engulfed Challenger. *Below, debris begins to rain down on the ocean.*

Right inset: A grim-faced Neil Armstrong listens to the testimony of one of the witnesses at the Rogers commission hearing.

following recommendations made by General Sam Phillips, who directed the Apollo program.

NASA was committing itself 'to expand human presence beyond the Earth into the solar system.' The NASA space station, due for launch in the 1990s, would be the first necessary step. It would provide a stepping stone to the rest of the solar system. From an expanded space station, an orbital service center would be constructed where lunar and planetary spacecraft could be built, launched and maintained.

The way ahead into the solar system, said Dr Fletcher, had been signposted in the report prepared in 1986 by the National Commission on Space, chaired by former NASA Administrator Tom Paine. The report was entitled 'Pioneering the Space Frontier: Our Next 50 Years'. It looked to a manned lunar outpost by the year 2004, and a full-scale manufacturing facility there by 2017. In that year too a manned outpost could be set up on Mars, leading to a viable base on the planet a decade later.

Concluding, Dr Fletcher said that he had asked the first American woman in space, Dr Sally Ride, to prepare a report, 'which will identify what we need to do to build the foundations of a revitalized United States space program.'

It was heady stuff, which reflected the determination and commitment of NASA to 'shake off the shadow of *Challenger*' and get back on track.

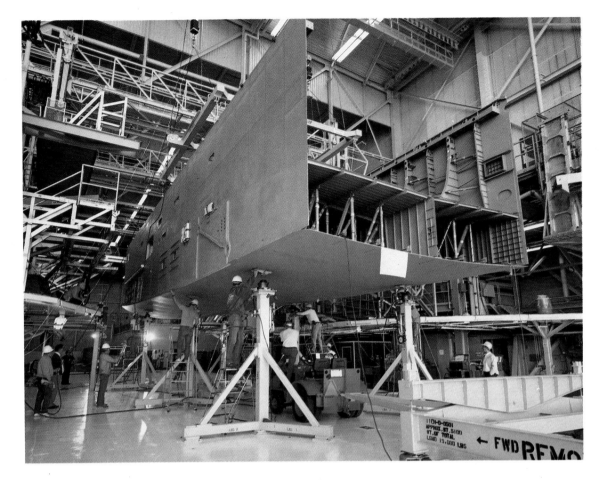

At Rockwell International's Palmdale plant, work on NASA's replacement orbiter, designation OV-105, gets underway. Workers are here jacking up the pre-assembled mid-fuselage section.

Meanwhile, up in orbit

While NASA agonized and had its dirty linen washed in public, major space rival Russia continued its formidable advance into the cosmos. Within days of *Challenger*'s demise, it launched its next-generation space station, Mir (meaning 'peace'). Some 46 ft (14 meters) long and with a maximum diameter of about 13 ft (4 meters), Mir is much the same size as its predecessor, Salyut 7, which has been in orbit since 1982 and which has served as home for some cosmonauts for up to 8 months.

Mir differs from Salyut in that it is outfitted primarily as living quarters and not as a workshop as well. It is divided into separate small cabins, which each contain a chair, desk and sleeping bag. The idea is that Mir serves as a base unit, which is capable of expansion into a much larger space station complex by the addition of science, experiment and workshop modules.

These modules can link up with the base unit via a spherical five-port docking module at one end or a single docking port at the other. The first science module to dock, in April 1987, was named Kvant (meaning 'quantum'). Kvant incorporates a pressurized laboratory and a scientific payload bay in which instruments such as X-ray telescopes are exposed to space.

At the year's end Mir cosmonaut Yuri Romanenko returned to Earth after 326 days in orbit, smashing all space duration records. Although he appeared reasonably fit, if bad-tempered on his return, Press reports later suggested he was a 'physical wreck' with severe calcium deficiency and muscle wastage, and suffering from a massive reduction in blood volume. The long sojourn in zero-G

had apparently taken its toll, and space scientists began to question the wisdom of keeping human beings in space for such long periods, given the present level of knowledge in space medicine.

A measure of Russia's dominance in space exploitation can be gauged from the world launch statistics for 1986, a year in which the number of launches in the West plummeted to an all-time low. For much of the year all of the West's major launch vehicles were grounded. They included not only the American shuttle, Delta, and Titan, but also Europe's Ariane launcher, which suffered upper-stage failure. And the launch score for the 12 months ending 31 December 1986 read: Russia 91, Rest of the World 12.

ELVs rule

With the exception of January's shuttle missions, all the launches had been made by unmanned ELVs (expendable launch vehicles), that is, by conventional multistage rockets. With the development of the shuttle NASA, and the military, had decided to phase out ELVs. The shuttle, they argued, could launch three or more satellites at a time. And the anticipated launch-every-two-weeks capability of a fleet of four shuttles would more than accommodate the satellite launch demands for the foreseeable future.

In practice, however, up until the time of the *Challenger* disaster, the shuttle had not even begun to realize its full potential. The maximum launch rate per year had been nine, in 1985. Also, the satellite launchings since the shuttle went commercial in November 1982 had not been 100 per cent successful, even though an astronaut crew was in attendance.

So even before the loss of *Challenger*, it was becoming clear that shuttles alone would be hard pressed to meet future satellite launch requirements – even for the United States market. After the loss, of course, there was no way that a reduced fleet of three shuttle orbiters could cope.

But the die had been cast; ELVs were on the way out. By mid-1986 just a handful of Titans, Atlases, Atlas-Centaurs and Deltas remained in the national inventory. So, to mix metaphors, NASA and the military, having put all their eggs in one basket, were caught with their pants down.

Back in production

To overcome the nationally embarassing shortfall in launch capability, contracts were hastily put out for a range of ELVs – to Martin Marietta, for example, for new Titans; to McDonnell Douglas for Deltas. And by August 1987 the first new Titans had begun to run off the production lie at Martin Marietta's Denver plant.

In the following October, NASA issued a mixed fleet manifest reflecting the switch from an all-shuttle to a mixed shuttle and ELV launch program through the 1990s. The Delta, for example, will be pressed into service to launch the Cosmic Background Explorer (COBE) in 1989, and the Roentgen satellite in 1990. Some payloads, though, have been purpose-built for the shuttle and can only be launched by it. They include the Hubble Space Telescope and the Galileo probe to Jupiter, both targeted for a 1989 launch.

ELVs have been ruled out also for launching hardware for the construction of the NASA space station. Using ELVs would greatly increase the astronauts' workload and the risk of accidents.

Up for grabs

The commercial side of NASA's launch program seems set to be phased out. With commitments to the military, the space station and space science taking precedence, there will be little room for private customers such as Western Union and the American Satellite Company. This will represent a considerable loss of revenue for the Administration. American customers will have to go elsewhere, as will foreign countries like Indonesia, Saudi Arabia, Mexico and Australia, all of whom have regularly launched with NASA.

Denied access to space via NASA, where will the customers go? In the long term, undoubtedly homespun rocket-launching companies will grow up, such as the already existing Transpace Carriers. In the short term customers may be forced overseas. A major contender is Europe's Arianespace, which operates the European Space Agency's Ariane launcher. It launches out of the favorably placed Kourou site in French Guiana, South America, close to the equator.

Above: A test-firing of the tractor rocket for a possible shuttle escape system. The rocket is designed to help pull the crew clear during the glide phase of landing.

Left: An astronaut demonstrates the crew escape position in the crew compartment trainer at the Johnson Space Center.

Russia has offered launch opportunities to foreign customers at favorable rates, though it is unlikely for political reasons that they will be taken up in the West. In the developing countries, however, it may be a different story. China too is seeking to exploit its growing expertise in space launchings commercially, and is offering offering its Long March rockets for international launches. Almost certainly Japan, already with an ambitious space program of its own, will also soon enter the commercial marketplace.

Space science in the 1990s

In one aspect of space exploration NASA has reigned supreme since the Space Age began – space science. Its very first satellite, Explorer 1, discovered the Van Allen radiation belts. In more recent years satellites such as Einstein and IRAS, and probes such as Viking and Voyager, have returned data that have revolutionized study of our solar system and indeed of the universe itself.

In 1987, while NASA was revamping its upcoming space science program in the wake of the *Challenger* disaster, 'old faithful' satellites and probes were continuing their robot exploration of space. The indefatigable Voyager 2 probe was well on the way to Neptune on schedule for an August 1989 encounter. The veteran International Ultraviolet Explorer (IUE) and Solar Max (Solar Maximum Mission) satellites were trained on the astronomical spectacle of the year – a supernova explosion in our galactic neighbor, the Large Magellanic Cloud. Solar Max, repaired in orbit on shuttle mission 41-C in 1984, acquired data that seemed to confirm the belief that the stuff the solar system is made of originated in a supernova explosion of a nearby star aeons ago.

In the fall of 1987 NASA announced new target dates for launching spacecraft that would spearhead space science investigations through the 1990s, assuming a resumption of shuttle flights in the summer of 1988. Two of the most critical payloads are the shuttle-borne Hubble Space Telescope (HST) and Galileo, both retargeted for 1989. Both had been ready for launch in 1986 within months of the *Challenger* disaster.

The $1.2 billion HST satellite carries the largest telescope ever to be deployed in space. Given its 7.9-ft (2.4-meter) diameter mirror and the crystal clarity of space, it will see much farther into the universe than we can from Earth and detect very much fainter objects. It will operate not only at visible light wavelengths, but also in the ultraviolet region of the spectrum at wavelengths that cannot penetrate the Earth's atmosphere. Its five major instrument packages include a Faint Object Camera developed by the European Space Agency (ESA).

The HST is designed for an extended lifetime. Its construction is modular, and it is designed for in-orbit servicing from the space shuttle. Operations control for the HST is at the Goddard Space Flight Center at Greenbelt, Maryland, while HST data will be analyzed at the Space Science Institute at nearby Baltimore.

Visiting Gaspra and Ida

Galileo, the first major planetary science payload since Voyager 2, is aiming for a rendezvous with giant planet Jupiter in 1995. But the launch constraints are critical, dictated by the celestial

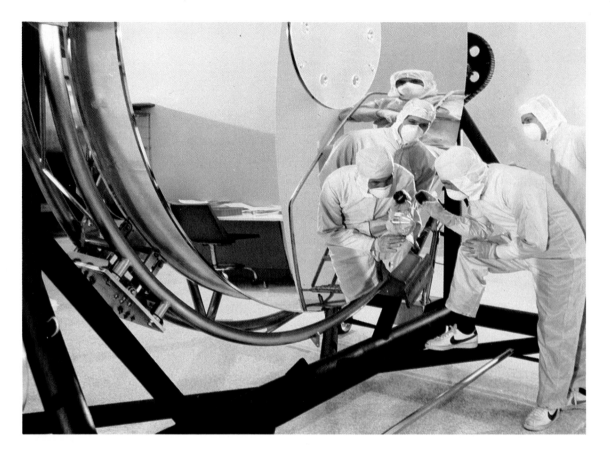

mechanics of the solar system. There is a launch window of just six weeks in the fall of 1989 when the mission can be attempted. If for any reason the launch window cannot be met, Galileo will have to wait a further 19 months for the next favorable planetary alignment.

The original itinerary for Galileo had to be radically modified when NASA banned the use of the liquid-fuel Centaur rocket as a booster for the probe. The hazards associated with having such a rocket in the payload bay of the shuttle were unacceptable. A less powerful solid-fuel booster was substituted, which demanded a more complex trajectory to the target planet.

Galileo will acquire the requisite speed for a suitable encounter by a series of gravity-assist maneuvers. In these it will use the sling-shot effect of planetary gravity to accelerate and redirect it to its destination. NASA has used the gravity-assist technique several times (for example, Mariner 10, Voyager 2), but never so ingeniously. After launch in October/November 1989, Galileo will coast towards Venus, loop round the planet in February 1990 and, speeding up, return to Earth the following December. It will then be flung out to the inner edge of the asteroid belt, where it will fly by asteroid Gaspra for a close encounter. After looping back to within 200 miles (300 km) of Earth again in December 1992, it will finally make its way towards Jupiter, via another asteroid, Ida.

In the summer of 1995 Galileo, now rapidly homing in on Jupiter, will let loose a spherical instrument probe designed to penetrate deeply into the thick Jovian atmosphere. By the year's end Galileo will be in orbit around Jupiter, where for two years it will transmit high-resolution images and data on the planet and its four large moons – Io, Europa, Ganymede and Callisto.

Left: Technicians at Perkin-Elmer's Optical Facilities at Wilton, Connecticut, examine the newly coated surface of the 94-inch (2.4-meter) primary mirror of the Hubble space telescope. They wear masks and special suits to prevent contamination of the pristine surface.

Right: In orbit high above the Earth's distorting atmosphere, the Hubble space telescope looks into the deepest reaches of space. It spies exploding and interacting galaxies, which spew out energy at an unbelievable rate at all wavelengths of the spectrum, and those baffling objects known as quasars, which appear to lie at the very edge of the known universe.

Also on the shuttle launch schedule for 1989 are the probe Magellan and the satellite Astro-1. Magellan, to be targeted on Venus, is equipped with an advanced imaging radar to map the planet from orbit. Astro-1 is an observatory for ultraviolet astronomy. A Delta rocket will launch Cosmic Background Explorer (COBE) in 1989. This satellite will measure, map and analyze the background radiation of the heavens, which is thought to have originated in the Big Bang that created the universe.

In 1990 the launch is scheduled of Ulysses, a joint NASA-ESA mission. NASA will provide the launch vehicle, the shuttle, and ESA the spacecraft. Ulysses will study regions of space never before investigated, above and below the plane of the ecliptic – the plane in which the planets revolve around the Sun. The probe will be launched into a trajectory that will use the sling-shot effect of Jupiter's gravity to direct it below, through and then above the ecliptic over both poles of the Sun.

Work is also well advanced on another interesting project to observe asteroids and comets, with a probe known as CRAF (Comet Rendezvous Asteroid Flyby). It is due for launch in 1993. CRAF will be launched into solar orbit to photograph asteroids and comets. It will also study a target comet (Tempel 2) over an extended period, and shoot a probe onto the comet. This will contain a miniature laboratory to analyze the 'dirty snowball' material the comet is made of.

'The next logical step'

In 1995, when CRAF will hopefully be encountering comet Tempel 2, NASA's major space effort will be directed towards the construction of its space station. This ambitious undertaking will increasingly dominate NASA's thinking and budget in the years ahead.

There are persuasive arguments in favor of a permanently manned space station. On the shuttle orbiter there is little room for scientific experiments, especially if the maximum crew of seven or eight is on board. Only on Spacelab flights is there room for meaningful science in the purpose-built space laboratory carried in the orbiter payload bay. But such flights are none too frequent and are of relatively short duration – up to about 10 days. They whet the scientific appetite rather than satisfy it. Only a permanent manned laboratory can give scientists and engineers the room and the time they require. The lengthy Skylab missions in the 1970s first demonstrated the benefits such a long-stay laboratory would bring.

Appearing before a House of Representatives Subcommittee on Space Science and Applications in February 1984, NASA Administrator James Beggs put it this way: 'To maximize the unique advantages provided by the environment of space, we need to establish a permanent presence that enables us to work in space full time. A space station is, I believe, the next logical step in space'. And only a few weeks earlier President Reagan had directed NASA to take that step 'within a decade'.

Many design concepts for space stations had already been suggested by aerospace companies such as Boeing, Rockwell and TRW, as well as by NASA centers such as Marshall and Johnson. It was now time to refine the concepts into a single practical entity.

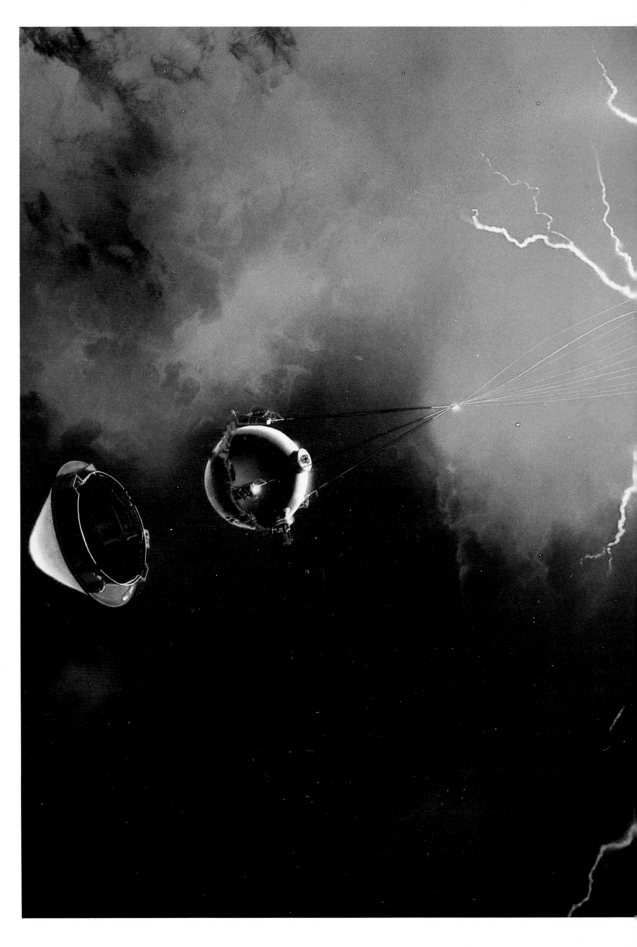

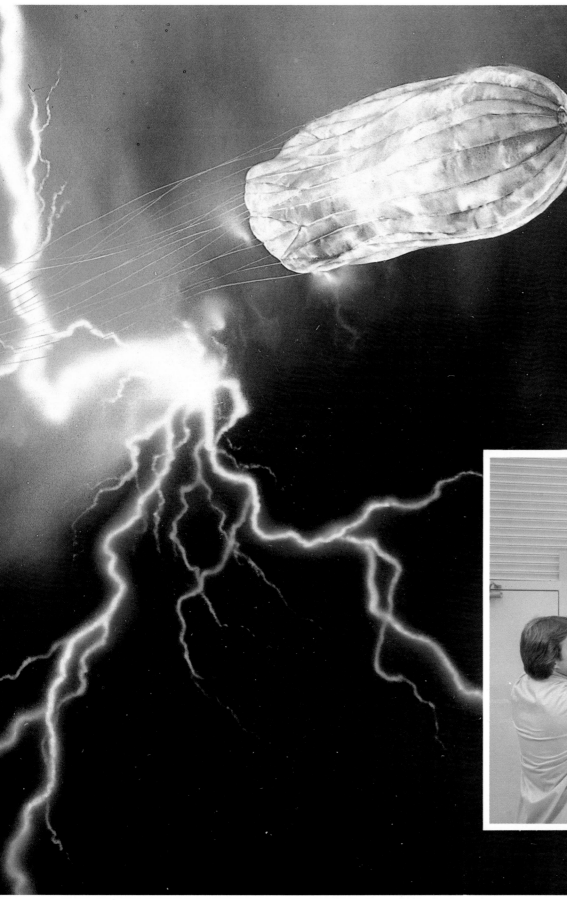

Detailed planning was allocated to four NASA centers – Johnson, Marshall, Lewis and Goddard. In addition Johnson was selected as the lead center responsible for the overall systems engineering of the station, the selection of the configuration and the integration of all elements into an operating system.

In December 1987 NASA announced the four companies selected for the first phase of space station construction – Boeing Aerospace, McDonnell Douglas, General Electric Company (GEC) and the Rocketdyne Division of Rockwell International. And the space station crept that little bit closer. By the month's end the urgent need to get something underway was underlined when cosmonaut Yuri Romanenko returned to Earth after a 326-day stint in the Russian space station Mir.

A modular design

The space station will be made up of an assembly of different units. They will be ferried into orbit by the shuttle, and put together *in situ* at an altitude of about 500 km (300 miles) by a team of astronaut-engineers flying in MMUs, with the assistance of the robot arms of one or more orbiters.

Left: Galileo's entry probe plunges into Jupiter's thick and stormy atmosphere. The heat shield separates as a parachute opens to brake the probe's descent.

Below: The entry probe and its heat shield during proving tests. Scientists expect the probe to report on conditions in the Jovian atmosphere for about 75 minutes.

As many as 20 shuttle flights may be required to lift into orbit the 200 tons of space station hardware. The first one, on current planning, should lift-off in 1994, and be followed at intervals by the others over a two-to-three-year period. The station could be operational by 1997. Budget cuts, however, could impact upon these dates.

In the currently planned configuration, the space station will consist of a number of cylindrical modules attached to a girder skeleton, or 'keel'. The crew of eight astronauts will have their living quarters in a habitation module. They will work in a number of laboratory modules. The modules will be of similar basic design but outfitted in different ways for their different roles. They will be some 50 ft (15 meters) long and about 15 ft (4.5 meters) in diameter, and have an 'up-and-down' (that is, ceiling and floor) orientation for crew comfort. The modules will have a multilayer aluminum hull, protected from meteoroid impact by a so-called bumper shield.

The modules will be linked each end to cylindrical chambers, or nodes, which will provide access between them. The nodes will incorporate docking interfaces for visiting shuttles and for logistics modules. These will be units for storing and transferring supplies.

Electrical power will be generated by large arrays of solar cells, protruding like paddles from the station skeleton. The skeleton will eventually carry pallets of instruments open to the space

Above: Orbital construction work will be made easier by the use of automated machines like this beam builder, developed at the Marshall Space Center.

Right: On shuttle mission 61-B in November 1985 space spiderman Jerry Ross makes short work of assembling the ACCESS trussed beams.

environment and operated remotely from inside the station or even from Earth. Other instruments will be mounted on free-flying platforms located away from the station, most of which will be unmanned. Some, however, will be equipped with life-support facilities for occasional habitation.

In the space station era, the shuttle will act as both taxi and truck. It may eventually visit the station only four times a year, to change crews after their regular 90-day tour of duty and replenish supplies. Transportation in orbit between the space station and free-flying platforms, will eventually be provided by an in-orbit shuttle craft called an orbital transfer vehicle (OTV). Crews in the OTV will provide in-orbit maintenance for the space station itself, and also act as service engineers to repair and refuel orbiting satellites.

An international effort

NASA will shoulder the main responsibility for space station construction and will provide much of the hardware. But other countries are involved as well. Japan, for example, will build a laboratory module, dubbed Jem. Canada will supply a mobile servicing center, which will provide support for the initial assembly and construction of the station, and for its future expansion. The center will incorporate advanced robotic arms like the Canadian-designed remote manipulator arm of the shuttle orbiter.

The biggest international contribution, however, comes from Europe, through the European Space Agency (ESA). ESA's multifaceted space station program is known as Columbus. It includes a pressurized laboratory module, which will interface with those of the United States and Japan. It also includes a number of instrument-carrying platforms placed in suitable orbits. One, the so-called man-tended free-flyer, will be a free-flying pressurized laboratory. This will be launched by ESA's new Ariane 5 launch vehicle, and be tended by astronauts from Hermes, ESA's proposed shuttle. Ariane and Hermes will also probably be involved in the launch and servicing of the other platforms.

The Orient Express

A mix of expendable launch vehicles and shuttles will serve American space flight needs into the space station era. But it is far from ideal, representing 'old technology'. The future, many believe, lies in the advanced hypersonic space plane. President Reagan first drew attention to the development of an American space plane in his 'State of the Union' address to a joint session of Congress in February 1986. It served as a timely morale booster just days after the stunning tragedy of the *Challenger* disaster.

The space plane will both take off from and land on a runway. It will be a single-staged vehicle requiring no external tank or boosters like the shuttle. it will be capable of accelerating to Mach 25 (25 times the speed of sound) to achieve orbital flight. 'Airline' versions, which would spend much of their flight time above the atmosphere, will be able to whisk passengers between New York and Tokyo, say, in just two hours. For this reason the American space plane has been dubbed the 'Orient Express'.

The official name for the plane is the National Aerospace Plane

(NASP). The initial objective of NASP research is to develop an experimental craft that will serve as a test-bed for the real thing. The design of the experimental vehicle, designated the X-30, will be chosen from competing designs advanced by General Dynamics, McDonnell Douglas and Rockwell International. The engine to power the plane will be designed either by Rocketdyne or Pratt & Whitney.

The X designation harks back to the experimental X-15 rocket plane program of the 1950s and '60s, which spearheaded hypersonic aerospace research at speeds above Mach 5. The X-15 plane carried some pilots above 50 miles (80 km), so high that they were awarded astronauts' 'wings'. Among X-15 pilots were two names that were to feature in later aerospace exploits – Neil Armstrong, the first moonwalker, and Joe Engle, who commanded the second shuttle flight.

The scram jet

The most revolutionary feature of the space plane will be its engines. A conventional launch vehicle has pure rocket engines, which carry their own supply of oxygen to burn their fuel. But the liquid oxygen generally used is heavy, accounting in the space shuttle, for example, for nearly three-fourths of the lift-off weight. So the idea in the space plane is for the engines to extract oxygen from the air during flight through the atmosphere. They would switch over to on-board liquid oxygen for the final push into orbit from the outer fringes of the atmosphere where the air is too thin to contain enough oxygen.

Astronauts will make widespread use of robot handling devices when they become involved in construction work up in orbit. Here they are using a robotic arm to position a connecting node for attachment to the existing station. This will allow another pressurized module to be attached.

A conventional jet engine, which also extracts oxygen from the atmosphere, is not suitable for speeds beyond about Mach 3. Above Mach 3 an engine can use the ramming effect of a plane's high forward speed to compress air in the combustion chamber. This is the principle behind the ramjet. But the ordinary ramjet cannot work at speeds beyond about Mach 6.

For higher speeds a special type of ramjet is needed, known as a scramjet, or supersonic combustion ramjet. This is designed to work with supersonic air flows through the combustion chamber. Experimental scramjets have already been tested at speeds beyond Mach 7, but much research needs to be done before they become a practical proposition.

The other major problem for the space plane is coping with the high temperatures generated by flight through the atmosphere. On the shuttle the problem is solved by 'passive' cooling — covering the airframe with tiles and other insulating materials, which radiate away most of the heat load. The space plane, however, will spend more time travelling through the atmosphere and be subject to greater heat stress. To cope with this, the plane will need to have an 'active' cooling system, especially in the nose, the leading edges of the wings and the engine. Active cooling will

Above: Close-up of a shuttle ferry docking on one of its periodic visits to the space station. It carries a new crew and fresh supplies.

Right: The baseline space station configuration comprises four pressurized modules, two supplied by the United States, one by ESA and one by Japan.

probably involve the pumping of liquid hydrogen fuel through ducts in the problem areas. This is the technique currently used for cooling conventional rocket engine nozzles.

The rivals

By the turn of the century American shuttles and X-30 space planes will not be alone in the skies. Shuttles and space planes from Russia, Europe and Japan may be up there with them. The Russians have long been developing their own space shuttle transportation system. It parallels the American shuttle in having an orbiter lifted off the launch pad by booster rockets, and returning to land on a runway.

C. BENNETT

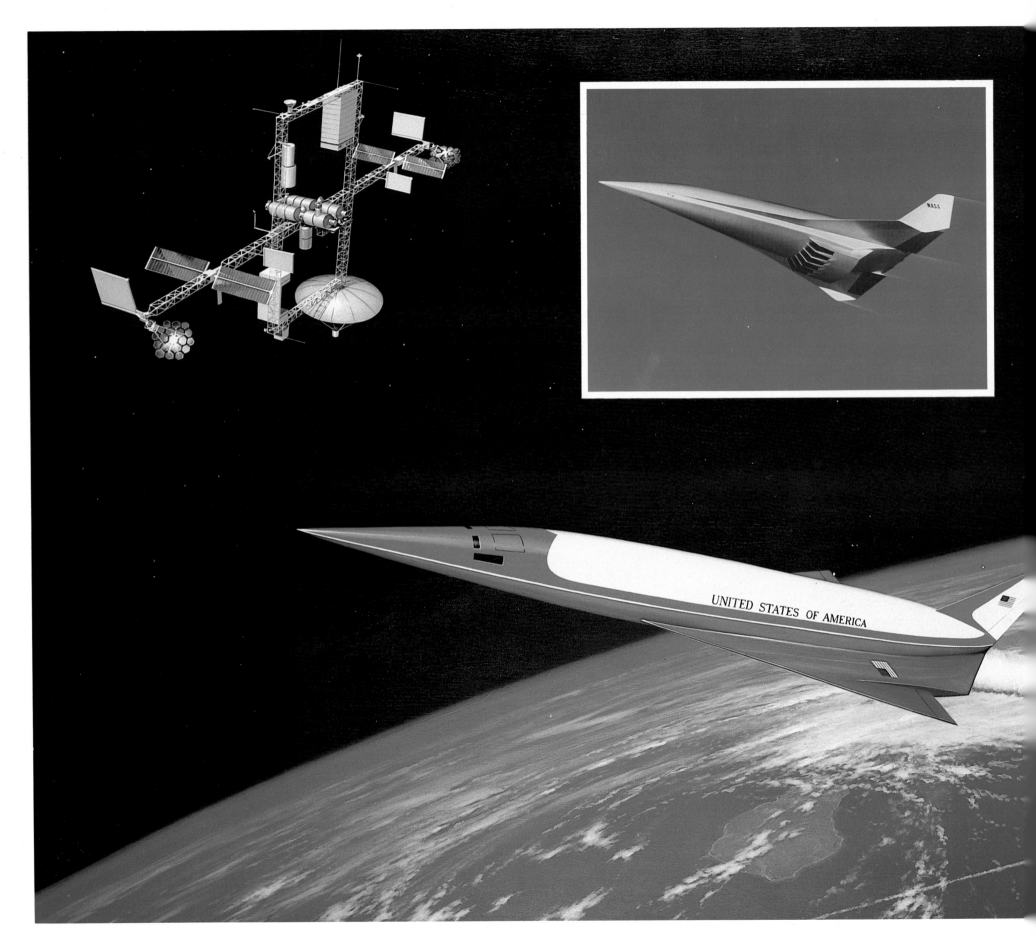

design features a revolutionary air-breathing rocket engine, invented by Alan Bond. This again reduces the amount of liquid oxygen needed on board, making for a lighter vehicle.

Return to the Moon

With the construction of the NASA international space station and the development of new shuttle and space planes, the decade of the 1990s promises to be one of the most exciting of the Space Age. That decade should also see the foundations laid for the 'expansion of human presence' into the solar system, as proposed in the Paine report of 1986 (see page 231) and the Ride report of 1987.

Dr Sally Ride, who has since left the astronaut corps, was lead member of a team charged by NASA to chart the future direction of the American space program, with one eye on the visions of Paine and the other on the realities of expedience, including a limited budget.

In her report Ride called for a steady growth in space exploration rather than for one-off spectaculars like Apollo. She concurred with Paine in recommending an early return to the Moon, and eventually a manned expedition to Mars. Closer to home, she suggested that increased resources be devoted to study of our own planet – spaceship Earth. For we still do not know enough about the Earth as an integrated system, about how the various elements – cloud cover, rainfall, vegetation, chlorophyll, ozone, carbon dioxide, and so on – interact.

Our return to the Moon should, Ride recommends, proceed in three phases in an evolutionary, not a revolutionary manner. Phase I, beginning in the 1990s, would focus on the robot exploration of our nearest neighbor in space. A lunar geoscience probe would map the surface from orbit and determine its geological make-up. Robot roving craft may then be landed to continue the exploration process.

Left: Accelerating to 25 times the speed of sound, a US space plane leaves the atmosphere behind as it heads for a rendezvous with the space station.

Top left inset: Another design proposed for the National Aerospace Plane, which will draw on the technical know-how acquired during the flights of the legendary X-15 (above).

But the orbiter will not have rocket engines like the American orbiters. It will ride atop the powerful rocket Energia, which made its impressive first flight in June 1987. The orbiter may have auxiliary jet engines to give it more maneuverability when returning to land. There appear to be two versions of the Russian shuttle, sometimes called Kosmolyot. The smaller one (Mischka) is probably a scale model of the other.

One European shuttle is called Hermes. Evolving from a French design, it will be launched by a more powerful version of Europe's heavy launching vehicle, Ariane. Japan has a similar plan to launch a shuttle craft called Hope (H-II orbiting plane) on its H-II launch rocket.

Germany has come up with a rather different concept, more in line with original NASA shuttle thinking. Known as the Sanger space plane after innovating German aerospace pioneer Eugen Sanger, it envisages a shuttle craft (Horus) carried piggy back to high altitude by a hypersonic transport plane. The shuttle would then boost itself into orbit, while its Earthbound carrier would return to base.

More directly parallel to the American X-30 space plane is Britain's Hotol (horizontal take-off and landing) vehicle. It is a single-stage launcher designed to take-off from and land on a runway and is intended primarily as a low-cost manned satellite launcher. Developed by British Aerospace and Rolls-Royce, the

At home on the Moon

Phase II, beginning in about the year 2000, would see astronauts return to the lunar surface after an absence of nearly 30 years to lay the foundations of a permanent outpost. The astronauts would be transported to and from the Moon in a lunar transfer vehicle, which would be assembled in Earth orbit at the space station. They would descend and ascend from the surface in a two-part landing module, not unlike the Apollo lunar module. On each trip the astronauts would ferry items of equipment to build the outpost – habitation modules, constructional materials, scientific instruments, roving vehicles, machinery, and a pilot plant for extracting oxygen from the lunar soil.

By 2005 the outpost would have grown into a base capable of permanent occupation. Phase III will have begun. Within five years or so the base would become a major scientific research facility, where scientists would exploit the Moon's low gravity (one-sixth Earth's) and vacuum. Astronomers would applaud the lack of atmosphere and the clarity of the dark skies. Lunar miners and metallurgists would extract and process valuable ores into constructional metals for ever more ambitious construction projects on the Moon itself and maybe in lunar orbit.

To the plains of Mars

Exploring and prospecting the Moon, establishing the Moon base and learning to exploit the Moon's resources would provide the experience and expertise necessary to extend the human presence farther afield. After the Moon, the next 'logical step' would take us to Mars, the only other planet in our solar system where human beings could land and live to tell the tale.

The exploration of the well-named Red Planet would broadly follow the same pattern as that of the Moon – investigation of the Martian landscape from orbit, followed by robot exploration and sampling of the surface. The first stage is already in prospect. Approval has been given for a Mars Observer probe that will conduct a two-year study from Martian orbit. It should be ready for launch in the early 1990s. It could be followed in the late 1990s by a Mars sample and return mission.

The robot explorers would pave the way for a manned mission some time in the second or third decade of the next century. At this distance in time, the logistics of a manned Mars mission are guesswork. But the problems are formidable. Even though Mars is a neighbor in Space, it still lies 35 million miles (56 million km) away even when most favorably placed. The minimum time for a round trip to the planet would be no less than two years.

Requirements: Build a spacecraft that will function reliably, given

Above: Photographs taken by the Apollo astronauts reveal the 'magnificent desolation' of the Moon. The rock and soil samples they returned show the Moon to be a rich source of many metals, which could readily be extracted at mining camps (left).

Right: Many people, both inside and outside NASA, still have their eyes on the Moon and predict a return there by the turn of the century. The lunar scene illustrated here shows mining activities. The mining and in-situ processing of lunar ores could provide the raw materials for constructing a permanent Moon base. Even oxygen could be extracted from the oxide ores that abound on the Moon.

that, after the first few days, it will be quite beyond help from Earth. Devise a life-support system that will keep an astronaut crew alive and provisioned, and protect them from known hazards such as cosmic radiation, solar flares and meteoroid particles. Keep the crew fit enough during the long weightless interplanetary cruise phase (98 per cent of mission time), so that they will physically be able to cope with the demanding surface exploration of Mars under gravity that will follow. Stop them becoming bored.

It will be difficult, though certainly not impossible, to meet these requirements on the proposed time scale. Perhaps a bigger stumbling block will be the cost, which could soar as high as $100 billion. But is this so high? In terms of today's money, the Apollo project cost upwards of $75 billion.

However, is it sensible to go-it-alone on such a journey into the unknown? Why not share the costs, research, risks, yes, and the glory, with other spacefaring nations, as is already happening with the space station. The Russians in particular are already committed to in-depth exploration of Mars, beginning in 1988 with an exploratory mission to its diminutive moon Phobos. NASA Administrator James Beggs suggested a joint American/Russian mission to Mars as early as 1985, on the 10th anniversary of the joint American/Russian ASTP mission. Surely this is the way ahead into the cosmos, through cooperation, not competition.

Right: A well-established Mars base in the third decade of the the next century. Water has proved not to be a problem, since reservoirs have been found deep inside the crust. Food is no problem either, being grown inside large greenhouse complexes. For long-distance transportation a Mars plane has been developed which can exploit the rarefied atmosphere.

NASA Milestones

1958

● Wernher von Braun's Army Ballistic Missile Agency (ABMA) team launches the first American artificial satellite, Explorer 1, from Cape Canaveral on 31 January, less than four months after Russia pioneered space flight with Sputnik 1 (launched 4 October 1957).

● The National Aeronautics and Space Administration (NASA) comes into being on 1 October.

1959

● The first seven US astronauts are selected in April for the Project Mercury manned spaceflight program.

● Pioneer 4 probe passes within 37,000 miles (60,000 km) of the Moon on 4 March and goes into solar orbit.

● On 28 May two monkeys called Able and Baker are recovered alive and well after a flight in a Jupiter rocket.

1960

● The first weather satellite, Tiros 1, is launched on 1 April, and the first passive communications satellite, Echo 1, on 12 August. An aluminized plastic balloon some 100 feet (30 meters) in diameter, Echo 1 is easily visible in the night sky.

1961

● Work begins on the space launch site on Merritt Island, later to become the Kennedy Space Center.

● Chimpanzee Ham has a successful suborbital flight in a Mercury capsule on 31 January.

● Alan Shepard in Mercury capsule *Freedom 7* emulates Ham's feat with a 15-minute suborbital flight on 5 May, less than a month after Soviet cosmonaut Yuri Gagarin pioneered manned orbital flight (in Vostok 1 on 12 April).

● President Kennedy on 25 May commits the US to the goal of putting an American on the Moon 'before the decade is out'.

● Virgil Grissom in *Liberty Bell 7* becomes the second US suborbital astronaut on 21 July.

● First test of the Saturn I heavy launch vehicle occurs on 27 October.

● Chimp Enos makes a two-orbit flight in a Mercury capsule, and is recovered in good shape on 29 November.

1962

● John Glenn in *Friendship 7* becomes the first orbiting US astronaut, making three orbits on 20 February.

● First orbiting solar observatory (OSO-1) starts detailed survey of the Sun from space on 7 March.

● Ariel 1, the first international satellite (with Britain), launched to investigate ionosphere on 26 April.

● On 24 May Scott Carpenter in *Aurora 7* also flies into space on a three-orbit mission.

● Communications satellite Telstar 1 inaugurates live transatlantic TV transmissions on 10 July.

● Walter Schirra pilots Mercury spacecraft *Sigma 7* for six orbits on 3 October.

● Mariner 2 becomes the first successful interplanetary craft in December by reporting conditions on Venus.

1963
● With a flight of 22 orbits, Gordon Cooper in *Faith 7* completes the Mercury program, 15-16 May.
● Syncom 2 becomes the first operational geostationary communications satellite on 26 July.

1964
● On 31 July Moon probe Ranger 7 hits its target, taking thousands of close-up photographs before impact.
● SERT 1, launched 20 July, tests ion engine in space.

1965
● John Young begins his illustrious space career when he and fellow astronaut Virgil Grissom make the first Gemini two-man flight (Gemini 3) on 23 March.
● Early Bird (Intelsat 1) becomes the first commercial communications satellite when it is launched on 6 April.
● Between 3-7 June, James McDivitt and Edward White in Gemini 4 complete a record (for the US) 62 orbits in space. On 3

June White makes the first American spacewalk, two-and-a-half months after cosmonaut Alexei Leonov's pioneered spacewalking (on 18 March).
● Mariner 4 sends back the first close-up pictures of Mars on 14 July from 6000 miles (9500 km).
● Gordon Cooper and Charles Conrad endure a 120-orbit, near 8-day journey in Gemini 5 (21-29 August).
● Frank Borman and James Lovell, in Gemini 7 (4-18 December) rendezvous in orbit with Walter Schirra and Thomas Stafford in Gemini 6A (15-16 December). Gemini 7 sets world record for space endurance of 206 orbits and 330 hours.

1966
● On 16 March Neil Armstrong and David Scott in Gemini 8 achieve first docking in space, linking up briefly with an unmanned Agena target vehicle.
● On 1 June lunar probe Surveyor 1 soft-lands on the Ocean of Storms, four months after Russian probe Luna 9 landed an instrument capsule. Surveyor sends back thousands of good-quality pictures.
● Between 3-6 June Thomas Stafford and Eugene Cernan fly into orbit in Gemini 9, Cernan carrying out a 2-hour EVA.

● John Young makes his second Gemini flight, this time with Michael Collins in Gemini 10 (18-21 July). After docking with an Agena, they use its engine to boost them to a record altitude for manned spacecraft of some 475 miles (760 km).

● Lunar Orbiter 1 enters lunar orbit on 14 August, becoming the Moon's second satellite after Russia's Luna 10 (3 April).

● Charles Conrad returns to orbit with Richard Gordon in Gemini 11 (12-15 September). They repeat rendezvous and docking maneuvers with an Agena target, which boosts them to a height of nearly 850 miles (1400 km).

● The Gemini missions end with Gemini 12 (November 11-15), during which Edwin Aldrin completes a record 5½-hour EVA.

1967

● Tragedy strikes at the Cape as the first crew assigned for a lunar landing perish in a flash fire during training on 27 January. They are Virgil Grissom, Edward White and Roger Chaffee.

● On 9 November the gigantic Saturn V rocket, designed to speed the Apollo spacecraft to the Moon, is successfully launched from Merritt Island's Complex 39.

1968

● The first manned Apollo flight (Apollo 7) lifts off on 11 October for a flight lasting 11 days. On board are Walter Schirra, Donn Eisele and Walter Cunningham.

● The first successful large-scale orbiting astronomical observatory (OAO-2) is launched on 7 December to investigate space in the ultraviolet region of the spectrum.

● The most ambitious flight in Earth's history begins on 21 December as the Apollo 8 command and service module (CSM) lifts off to attempt a circumnavigation of the Moon. During the six-day trip, the crew of Frank Borman, James Lovell and William Anders, orbit the Moon 10 times.

1969

● The whole Apollo spacecraft, CSM and lunar module (LM), is tested for the first time in Earth orbit on the Apollo 9 mission (3-13 March), with James McDivitt, David Scott and Russell Schweickart aboard.

● Thomas Stafford, John Young and Eugene Cernan, in Apollo 10 (18-26 May) fly to the Moon and practice LM separation and re-docking in lunar orbit, in a full dress rehearsal for the first Moon landing.

● On 16 July Neil Armstrong, Edwin Aldrin and Michael Collins blast off in Apollo 11, bound for a Moon landing. On 20 July Armstrong sets foot on the Moon's Sea of Tranquillity, followed shortly afterwards by Aldrin; they spend some 2½ hours exploring. The astronauts return to Earth triumphantly on 24 July.

● Mariner 6 and 7 probes fly by Mars (31 July and 5 August) and send back the first detailed pictures of the Red Planet.

● The Apollo 12 mission (14-24 November) succeeds in making the second Moon landing on the Ocean of Storms. Charles Conrad and Alan Bean do the Moon-walking with two EVAs totaling nearly 8 hours, while Richard Gordon remains in lunar orbit.

Left: A historic moment indeed is captured here at Houston mission control on 20 July 1969. Controllers supervize the first lunar EVA by Apollo 11 astronauts Neil Armstrong and Edwin Aldrin, seen on the TV screen.

Below: The huge satellite ATS-6 is seen being tested in the Space Environment Simulation Laboratory at the Johnson Space Center before its launch in 1974. It pioneered satellite broadcasting to small ground receivers.

1970

● Unlucky Apollo 13 astronauts James Lovell, John Swigert and Fred Haise nearly lose their lives when their CSM is crippled en route to attempt the third Moon landing. But they make it safely back to Earth after six days, on April 17, after using the LM as a liferaft.

1971

● The next Apollo mission (Apollo 14, 31 January-9 February) reaches the Moon, lands in the Fra Mauro region and returns safely. While Stuart Roosa orbits above, Alan Shepard and Edgar Mitchell use a handcart to carry their equipment on two EVAs of some 9 hours. It was the second wheeled 'vehicle' on the Moon, after the Russian robot explorer Lunokhod 1, which had landed the previous November.
● Apollo 15 astronauts David Scott and James Irwin have even better transport in the electrically propelled lunar rover during their mission (26 July-7 August) to Hadley Rille. Their three EVAs total more than 18 hours. CSM commander is Alfred Worden.
● Mariner 9 becomes the first artificial satellite of Mars when it enters orbit on 13 November, and comprehensively maps the planet.

1972

● On 5 January President Nixon approves the development of a space shuttle.
● Pioneer 10 begins a 21-month journey to Jupiter on 3 March.
● The penultimate Apollo mission (Apollo 16, 16-27 April) sees John Young making his fourth space flight. With Charles Duke, he explores the Cayley Plains region of the lunar surface, leaving Thomas Mattingly in the CSM, in three EVAs totaling some 20 hours.
● Rocket designer extraordinary Wernher von Braun retires from NASA on 1 July.
● The first Earth Resources Technology Satellite (ERTS-1), later named Landsat 1, is launched from the Western Test Range on 23 July to provide the first comprehensive satellite mapping of the Earth.
● OAO-3 (Copernicus) starts investigation of the heavens at X-ray wavelengths on 21 August. It gathers data on a possible black hole, Cygnus X-1.
● The Apollo project draws to a triumphant conclusion as Apollo 17 (7-19 December) makes the sixth lunar landing at Taurus-Littrow. During three EVAs Eugene Cernan and Harrison Schmitt roam the surface for a record 22 hours. Ronald Evans is CSM commander.

1973

● The launch of the experimental space laboratory Skylab nearly ends in disaster when a solar panel and insulation are ripped off during the ascent on 14 May atop a Saturn V booster.
● The first Skylab crew are ferried to Skylab on 25 May, and succeed in making running repairs to it. The crew of Charles Conrad, Joseph Kerwin, and Paul Weitz return home on 22 June after 28 days.

● The following month the next Skylab crew, Alan Bean, Jack Lousma and Owen Garriott, soar into orbit (on 28 July). They remain aloft for 59 days, returning on 25 September.
● The final Skylab crew lift off on 16 November for a record-breaking 84 days, by far the longest any Americans have remained in orbit. They are Gerald Carr, Edward Gibson, and William Pogue. They return on 8 February.

1974

● Mariner 10 probe encounters Venus on 5 February, then uses the planet's gravity to swing it into a trajectory to Mercury, taking the first close-up pictures of this planet on 29 March.
● The first geostationary weather satellite SMS-1 reaches orbit on 17 May.
● After traveling some 600 million miles (one billion kilometers). Pioneer 10 encounters Jupiter in December, sending back the first close-up pictures of the giant planet. It approaches to within 81,000 miles (130,000 km) of the cloud tops.

Above: The year 1972 saw the launch of Landsat 1 (formerly ERTS-1), whose electronic eyes spied details of the Earth's surface never seen before. This image covers the Luichow Peninsula in China.

● On 30 May the huge Applications Technology Satellite ATS-6 is launched, the most powerful communications satellite to date, capable of broadcasting to remote communities via small receiving dish antennas.

1975
● First international manned space flight begins on 15 July, when an Apollo and a Soyuz craft link up in orbit during the Apollo-Soyuz Test Project (ASTP). The crew are Vance Brand, Thomas Stafford and Donald Slayton (US), and Alexei Leonov and Valery Kubasov (USSR),
● Viking probes set off on 20 August (Viking 1) and 9 September (Viking 2) bound for a landing on Mars.
● First geostationary operational environmental satellite (GOES-1) launched for NOAA on 16 October.

1976
● Viking probes land on Mars: Viking 1 lands at Chryse on 20 July, Viking 2 at Utopia on 3 September. They photograph the landscape and sample soil, but find no signs of life.
● Laser-reflecting satellite Lageos launched on 4 May to help study of Earth shape and crustal movements. Will remain in orbit for 10 million years.

1977
● The first high-energy astronomy satellite (HEAO-1) launched on 12 August to map X-ray and gamma-ray sources in the heavens.
● The prototype space shuttle orbiter *Enterprise* makes the first free flight from a Boeing 747 carrier aircraft on 12 August, gliding to a runway landing from a height of 22,800 feet (6950 meters). Piloted by Fred Haise and Gordon Fullerton.
● Voyager 2 deep space probe launched on 20 August.
● Voyager 1 launched on 5 September in a faster trajectory than Voyager 2.
● *Enterprise* makes final approach and landing test on 26 October, verifying the aerodynamic performance of the shuttle orbiter design. Haise and Fullerton are again at the controls.
● European geostationary weather satellite Meteosat 1 launched for the European Space Agency (ESA) on 22 November.

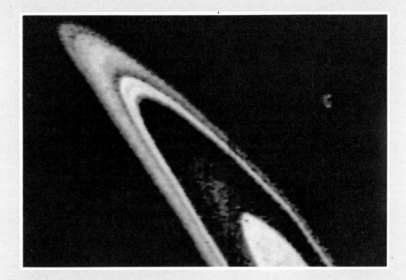

Left: In 1979 the Pioneer 11 deep space probe returned the first close-up pictures of Saturn and its enigmatic ring system.

1978
● International ultraviolet explorer (IUE) launched on 26 January to view the heavens at ultraviolet wavelengths. Cooperative project with the European Space Agency.
● Pioneer-Venus interplanetary probe launched to Venus on 20 May; goes into orbit around Venus on 4 December and conducts detailed survey of the planet in conjunction with entry probes into the atmosphere.
● Seasat 1 launched into Earth's orbit on 26 June to investigate the ocean surface, wave topography, surface wind speed, ocean temperature.
● First of a new series of weather satellites, Tiros N, goes into orbit on 13 October.

1979
● Voyager 1 makes closest approach to Jupiter (within 175,000 miles, 280,000 km) on 5 March, sending back a wealth of information and beautiful pictures of the giant planet and its satellites.
● Prototype orbiter *Enterprise,* mated with solid rocket boosters and external fuel tank, rolls out to the launch pad at the Kennedy Space Center for compatibility checks on 1 May.

Below: Viking 1 dropped a lander onto the Martian plain called Chryse in 1976. The lander sent back this panoramic view of the surrounding landscape.

- On 9 July Voyager 2 repeats its sister craft's triumph at Jupiter, flying within 405,000 miles (650,000 km).
- Skylab breaks up and its remains scatter over the Pacific and Western Australia when it drops from orbit on 11 July.
- Pioneer 11 becomes the first probe to travel to Saturn, approaching within 13,500 miles (21,500 km) on 1 September.
- The third high-energy astronomical observatory (HEAO-3) *Einstein* launched on 20 September. Surveys the heavens at X-ray wavelengths. Highly successful.

1980

- Solar maximum mission satellite (Solar Max) launched on 14 February but fails after 11 months. Later, it features in the first in-orbit satellite repair mission, on space shuttle mission 41-C, April 1984.
- Voyager 1 reaches Saturn and flies within 78,000 miles (126,000 km) of its cloud tops on 12 November. Sends back spectacular pictures of the rings and discovers many new moons.
- The first of the powerful Intelsat V communications satellites, with 12,000 voice circuits, is launched on 6 December.
- The first operational orbiter *Columbia* is rolled out to the launch pad on 29 December, looking to a launch in the following spring.

1981

- John Young and Robert Crippen fly *Columbia* into orbit on 12 April on the space shuttle's maiden mission (STS-1). It lasts for 36 orbits, 54 hours. Taking off from the Kennedy Space Center, it lands at the Edwards Air Force Base in California, on 14 April.
- On 25 August Voyager 1 makes its closest approach, 63,000 miles (101,000 km), to Saturn.
- *Columbia*, on the second shuttle mission (STS-2), roars into space again for a 36-orbit flight on 12 November, piloted by Joe Engle and Richard Truly. It is the first time any craft has returned to space. Crew test the remote manipulator system robot arm.

1982

- *Columbia* makes its third flight (STS-3) on 22 March, crewed by Jack Lousma and Gordon Fullerton. Lands at White Sands, New Mexico, after an 8-day flight.
- STS-4 begins on 27 June, with Thomas Mattingly and Henry Hartsfield flying *Columbia* on a 7-day mission, the final test mission.
- On 16 July the new-generation Landsat 4 Earth-resources satellite enters orbit.
- *Columbia* becomes operational on STS-5, lifting off on 11 November. It carries a record four-man crew, Vance Brand, Robert Overmyer, Joseph Allen and William Lenoir. They launch two communications satellites.

1983

- International infrared astronomy satellite IRAS launched on 25 January. Discovers comets, possible new solar systems and sees stars being born during its 10-month operational life.
- The maiden flight of the second operational orbiter *Challenger*, on STS-6, takes place on 4 April, with a crew of four, Paul Weitz,

Karol Bobko, Donald Petersen and Story Musgrave. During a 5-day flight the crew deploy the first tracking and data relay satellite (TDRS), and Petersen and Musgrave test the new shuttle spacesuit in the orbiter cargo bay.
- European X-ray satellite Exosat launched on 26 May.
- First American woman astronaut Sally Ride soars into orbit on *Challenger's* second flight (STS-7) on 24 June, along with Robert Crippen, Frederick Hauck, John Fabian and Norman Thagard, making a record five-person crew.
- On 13 June Pioneer 10 becomes the first probe to venture into interstellar space when it crosses the orbit of the outermost planet, then Neptune.
- On 30 August STS-8 launches spectacularly at night. The crew of *Challenger* includes Richard Truly, Daniel Brandenstein, Dale Gardner, Guion Bluford, William Thornton, and six rats. Landing on 5 September is also at night.
- On 1 October NASA celebrates a quarter-century of spectacular achievement.
- The first flight of Spacelab on the shuttle begins on 28 November. The orbiter is *Columbia*. The crew of six conduct over 70 experiments on a 10-day flight. Among them is German scientist Ulf Merbold, the first non-American to fly in the US space program. The other crew members are John Young, Brewster Shaw, Robert Parker, Owen Garriott and Byron Lichtenberg.

Below: Shuttle orbiter Challenger is pictured in orbit on its second mission (STS-7) against a backdrop of azure ocean some 180 miles (290 km) below.

1984

- *Challenger* blasts off for the 10th shuttle flight (41-B) on 3 February. Bruce McCandless test-flies the MMU (manned maneuvering unit) and becomes the first human satellite. Robert Stewart in the MMU later becomes the second. The other crew members are Vance Brand, Robert Gibson and Ron McNair. Two comsats, Westar VI and Palapa B2, are deployed, but fail to reach the correct orbits. *Challenger* becomes the first orbiter to land at the Kennedy Space Center on 11 February.
- The fifth Earth-resources satellite, Landsat 5, ascends into orbit on 1 March. Its thematic mapper can resolve details down to about 100 feet (30 meters).
- *Challenger's* next mission, 41-C, lifts off on 6 April with an ambitious program — the deployment of the long duration exposure facility (LDEF) and the capture and repair of the ailing Solar Max (Solar Maximum Mission) satellite. George Nelson flying the MMU failed to dock with it, leaving Terry Hart to catch it with the remote manipulator arm. Nelson and James van Hoften then effect the repairs. The other crew members are Robert Crippen and Francis (Dick) Scobee.
- The third orbiter, *Discovery*, becomes operational on 30 August on mission 41-D. Second US woman astronaut Judy Resnik helps extend a 105-foot (32-meter) solar panel mock-up. A record three satellites are deployed. Also on this, the 100th manned spaceflight, are Henry Hartsfield, Michael Coats, Richard Mullane, Stephen Hawley and Charles Walker.
- On 5 October Robert Crippen makes his fourth shuttle flight on 41-G. *Challenger's* record seven-person crew features two women astronauts. Sally Ride is making her second space flight, while Kathy Sullivan becomes the first US woman to spacewalk, with David Leestma. The other crew include Canadian Marc Garneau, oceanographer Paul Scully Power and Jon McBride.
- On *Discovery's* second flight (51-A), which started on 8 November, the crew launched two satellites and recovered two — the errant Westar VI and Palapa B2 from 41-B. The recovery was achieved thanks to spectacular spacewalking by Dale Gardner and Joseph Allen in MMUs, coupled with skilled operation of the remote manipulator arm by Anna Fisher. Fred Hauck and David Walker made up the crew.

1985

- A Department of Defense (DoD) payload is lifted into orbit by *Discovery* on mission 51-C, on 24 January. The crew on this, the shortest (3 days) operational shuttle flight to date, are Thomas Mattingly, Loren Shriver, Ellison Onizuka, James Buchli and Gary Payton.
- On 12 April, four years to the day after the first shuttle flight, flight 51-D leaves the launch pad. *Discovery's* crew includes Senator Jake Garn. The Leasat-3 comsat fails to work after deployment. It defies attempts by spacewalkers David Griggs and Jeffrey Hoffman with a 'flyswatting' device and Rhea Seddon with the manipulator arm to activate it. The rest of the flight crew were Karol Bobko, Donald Williams and Chas Walker.

- On flight 51-B, launched 29 April, *Challenger* carries Spacelab 3. The crew are Robert Overmyer, Frederick Gregory, Don Lind, Norman Thagard, William Thornton, Lodevijk van den Berg and Taylor Wang, together with two monkeys and 24 rats.
- Prince Sultan Abdul Aziz Al-Saud of Saudi Arabia is payload specialist on 51-G to supervize launch of Arabsat. The other 'foreigner' on board *Discovery* is Frenchman Patrick Baudry. The other crew members are Daniel Brandenstein, John Creighton, Steven Nagel, John Fabian, and 'the oldest woman in space' Shannon Lucid (42).
- *Challenger's* launch attempt on 12 July fails at T-3 seconds due to an engine malfunction. During launch on 29 July 'abort to orbit' is declared when No 1 engine shuts down due to faulty sensor. The mission (51-F) carries Spacelab 2, igloo and 2 pallets. The crew are Charles Fullerton, Roy Bridges, Story Musgrave, Anthony England, oldest man in space' Karl Henize (58), Loren Acton and John-David Bartoe.

Right: This image of Halley's comet was acquired by the astronomy satellite International Ultraviolet Explorer (IUE) on the last day of 1985, the year that saw the first return of the comet to our skies since 1911.

● 'Strongman' James van Hoften and Joe Engle, aided by manipulator arm handler Mike Lounge, capture and repair the errant Leasat-3 on flight 51-I (*Discovery*, launched 24 August). Also on board are Richard Covey and William Fisher, husband of Anna, who preceded him into space.

● Seven-year-old spacecraft ICE (International Cometary Explorer), formerly known as ISÈE (International Sun-Earth Explorer), reports on comet Giacobini-Zinner on 11 September. Two years earlier it had been ingeniously directed into a trajectory to the rendezvous by the sling-shot effect of lunar gravity.

● The maiden launch of fourth orbiter *Atlantis* takes place on 3 October. The mission (51-J) is classified, and faultless. The all-military crew are Karol Bobko, Ronald Grade, Robert Stewart, David Hilmers and William Pailes.

● The West German dedicated Spacelab D-1 mission (61-A) begins on 30 October. *Challenger* carries a record crew of eight. Payload operations controlled from German Space Operations Center at Oberpfaffhenhofen, near Munich. West Germany's Rheinhard Furrer and Ernst Messerschmid, and Wubbo Ockels of the Netherlands are the payload specialists. Other members of the crew are Henry Hartsfield, Steven Nagel, James Buchli, Guion Bluford and Bonnie Dunbar.

● *Atlantis* makes the second night launch of the shuttle on 26 November. Three comsats are deployed on this mission (51-B), including one for Mexico, represented on board by Rudolfo Neri Vela. Jerry Ross and Sherwood Spring take part in spectacular EVAs, in which they practice space-girder construction tehniques with EASE (Experimental Assembly of Structures in EVA) and ACCESS (Assembly Concept for Construction of Erectable Space Structures) equipment.

● On 16 December Jet Propulsion Laboratory scientists tune into signals from solar probe Pioneer 6 to celebrate its 20th anniversary in space. It is still working – not bad for a craft with a 6-month design life! Distance covered to date, some 12 billion miles (20 billion km). Late in the month the veteran IUE (International Ultraviolet Explorer) becomes the first probe to report back about Halley's comet. It detects the breakdown products of water, confirming that the comet contains water ice.

Left: On the West German dedicated Spacelab D-1 mission beginning 30 October 1985, Rheinhard Furrer bares his arm for bloodletting by a fellow crew member. Just in the picture is Wubbo Ockels, arm bared for the same purpose.

1986

● A much delayed *Columbia* lifts-off on 12 January on a troubled mission (61-C). Aboard is Congressman Bill Nelson, together with a regular astronaut crew of Robert Gibson, Charles Bolden, Franklin Chang-Diaz, Stephen Hawley, George Nelson and Robert Cenker.

● The much heralded 'year for space science' starts well, with Voyager 2 sending back spectacular images of Uranus and its moons from a distance of some 2 billion miles (3 billion km), over a few days before and after the close-encounter day of 24 January.

● NASA scientists are still in euphoria over Voyager results, when the tragedy that stunned the nation occurs at the Kennedy shuttle launch site on 28 January. *Challenger*, on its 10th and the shuttle's 25th mission (51-L), explodes 73 seconds after lift-off from pad 39B, in a fireball that smashes the craft to smithereens and kills the crew of seven – Dick Scobee, Mike Smith, Judy Resnik, Ellison Onizuka, Ron McNair, Greg Jarvis, and 'teacher in space' Christa McAuliffe. Subsequent investigation of the cause of the accident – the first in-flight disaster in US space history – points to failure of an 'O-ring' seal between segments of the right-hand solid rocket booster. The disaster leads to the immediate grounding of the shuttle fleet and an in-depth investigation of all NASA managerial and operational procedures.

● The USA space program suffers another blow on 18 April when a Titan 34D Air Force rocket, carrying a Big Bird spy satellite, blows up just after leaving the launch pad at Vandenberg Air Force Base.

● Unbelievably two weeks later (on 3 May), a Delta 3914 launch vehicle malfunctions and has to be destroyed 71 seconds after lift-off from the Cape.

● President Reagan approves the construction of a new orbiter to replace *Challenger*, at a cost of some $3 billion.

● On 5 September a Delta vehicle successfully launches a Strategic Defense Initiative (SDI) payload from the Cape.

● On 3 October NASA announces February 1988 as the target date for resuming shuttle flights, with orbiter *Disovery*.

● An Atlas-Centaur lifts-off successfully from the Cape on 4 December carrying a military comsat.

1987

● NASA announces the flight crew for the next shuttle flight (mission 26) – Fred Hauck, Richard Covey, John Lounge, George Nelson and David Hilmers.

● Ill-luck continues to dog NASA as an Atlas-Centaur launch vehicle goes out of control during launch from the Cape on 26 March and is destroyed.

● On 1 December NASA selects Boeing, McDonnell Douglas, General Electric Company and Rockwell as major contractors for the permanently manned space station scheduled for launch in the mid-1990s.

● Morton Thiokol test-fire the redesigned shuttle solid rocket booster (SRB) on 23 December. The 'O-ring' seal problem that apparently caused the *Challenger* disaster seems to have been solved. Subsequent inspection of the rocket, however, reveals a new problem, in the steering gear. The launch date of the next mission, already retargeted to June 1988, slips again, to August 1988 at the earliest.

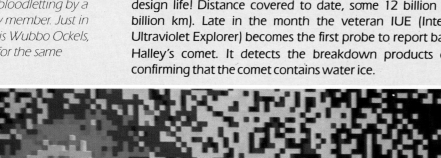

Index

Picture Credits

The author would like to extend his grateful thanks to his many friends at NASA centers across the nation for their immense help in the preparation of this book, and in particular for providing the majority of the photographs. He is also most grateful to the following individuals and organizations for additional pictures.
(B = Bottom, T = Top, L = Left, R = Right)
British Interplanetary Society 8, 12; **Eros Data Center** 179, 186, 188/9, 190; **Goddard Space Flight Center** 9; **Hughes Aircraft Company** 178T, 185; **Imperial War Museum** 10, 11; **Robin Kerrod** 16L, 22T, 83TR, 91 105 (inset), 106B, 107, 109T, 113, 149, 220, 221, 222, 223, 224, 225, 226B; **KPNO** 192, 213; **Novosti** 13T, 18/19, 24T; **RAE Farnborough** 183, 184, 231; **Rockwell International** 173; **US Naval Observatory** 177

Multimedia Publications (UK) Limited have endeavored to observe the legal requirements with regard to the suppliers of photographic material.